AF761134

Transforming Research Excellence

New Ideas from the Global South

Edited by Erika Kraemer-Mbula, Robert Tijssen,
Matthew L. Wallace and Robert McLean

AFRICAN MINDS

Published in 2020 by African Minds
4 Eccleston Place, Somerset West, 7130
Cape Town, South Africa
info@africanminds.org.za
www.africanminds.org.za

All contents of this document, unless specified otherwise, are licensed under a Creative Commons Attribution 4.0 International License.

ISBNs:
978-1-928502-06-7 Print
978-1-928502-07-4 e-book
978-1-928502-08-1 e-pub

Copies of this book are available for free download at www.africanminds.org.za

ORDERS
For orders from Africa:
African Minds
Email: info@africanminds.org.za

For orders from outside Africa:
African Books Collective
PO Box 721, Oxford OX1 9EN, UK
Email: orders@africanbookscollective.com

Contents

Preface and acknowledgements iv

01 Introduction | *Erika Kraemer-Mbula, Robert Tijssen, Matthew L. Wallace and Robert McLean* 1

Part 1 Theoretical and conceptual underpinnings

02 Redefining the concept of excellence in research with development in mind | *Judith Sutz* 19

03 The Republic of Science meets the Republics of Somewhere: Embedding scientific excellence in sub-Saharan Africa | *Joanna Chataway and Chux Daniels* 39

04 Re-valuing research excellence: From excellentism to responsible assessment | *Robert Tijssen* 59

05 Gender diversity and the transformation of research excellence | *Erika Kraemer-Mbula* 79

06 Research excellence is a neo-colonial agenda (and what might be done about it) | *Cameron Neylon* 92

Part 2 Research excellence in practice

07 Utility over excellence: Doing research in Indonesia | *Fajri Siregar* 119

08 Supporting research in Côte d'Ivoire: Processes for selecting and evaluating projects | *Annette Ouattara and Yaya Sangaré* 138

09 Sustaining research excellence and productivity with funding from development partners: The case of Makerere University | *Vincent A. Ssembatya* 147

10 Southern conceptions of research excellence | *Suneeta Singh and Falak Raza* 164

11 From perception to objectivity: How think tanks' search for credibility may lead to a more rigorous assessment of research quality | *Enrique Mendizabal* 178

Part 3 Striving for solutions

12 Exploring research evaluation from a sustainable development perspective | *Diego Chavarro* 203

13 Indicators for the assessment of excellence in developing countries | *Rodolfo Barrere* 219

14 Rethinking scholarly publishing: How new models can facilitate transparency, equity, efficiency and the impact of science | *Liz Allen and Elizabeth Marincola* 233

15 Research Quality Plus: Another way is possible | *Jean Lebel and Robert McLean* 248

16 Call to action: Transforming 'excellence' for the Global South and beyond | *Erika Kraemer-Mbula, Robert Tijssen, Matthew L. Wallace, Robert McLean, Liz Allen, Rodolfo Barrere, Joanna Chataway, Diego Chavarro, Chux Daniels, Jean Lebel, Elizabeth Marincola, Enrique Mendizabal, Cameron Neylon, Annette Ouattara, Falak Raza, Yaya Sangaré, Suneeta Singh, Fajri Siregar, Vincent A. Ssembatya and Judith Sutz* 259

About the authors 264

Index 271

Preface and acknowledgements

There is an increasing drive to steer funding towards research 'excellence' around the world. In the Global South, especially in low- and medium-income countries (LMICs), emerging granting councils face the challenge of supporting science that is both high quality and relevant to their own national priorities. However, recent scholarship has revealed that the notion of excellence is problematic in many, if not all, contexts. It is highly associated with subjective value judgements on disciplines, methodologies, and is closely linked to journal impact factors, H-index scores, sources of funding and university rankings, each of these being highly contested. In the Global South, many have explored to which degree scant research resources must be focused on development priorities. Given these developments, the time is ripe to fill the knowledge gap regarding research excellence in the developing world, providing balance to 'Global North-dominated' scholarship on this issue.

On a more practical level, initiatives such as the Science Granting Councils Initiative (SGCI) in sub-Saharan Africa have revealed pressures on research organisations in LMICs to demonstrate competitiveness in a global research space, and demonstrate that research is 'as good' as that which is done elsewhere. Partially driven by the same spirit of accountability and a desire to build capacity for 'world-class' science, external donors are increasingly pushing for their funds to go towards 'excellent' research. In both cases, the issue of quality and accountability cannot be ignored, as many governments are weighing the benefits of allocating larger budgets to scientific research. However, they are generally poorly equipped to evaluate research quality and excellence, and to use this evaluative evidence to manage the tensions between national research capacity (and capacity-building) issues, local relevance and demand for research, and various types of quality standards. This speaks to the need for more context-specific quantitative and qualitative indicators to assess and measure research quality, more robust methods for conducting research evaluation, as well as well-developed modalities and programme designs for supporting research.

The ideas in this book emerge from various sources. Our initial quest to learn more about 'research excellence in the Global South' arose from the SGCI. Beginning in 2015, the SGCI is a multi-funder initiative that aims to strengthen the capacities of 15 science granting councils (SGCs) in sub-Saharan Africa in order to support research and evidence-based policies that will contribute to economic and social development. It is funded and managed by Canada's International Development Research Centre (IDRC), the UK's Department for International Development (DFID), the National Research Foundation (NRF) (South Africa) and, since 2018, the Swedish International Development Agency (SIDA). It is guided by the priorities of the 15 granting agencies who, in 2016, sought to explore the notion of research excellence in greater depth, leading to a report by Erika Kraemer-Mbula and Robert Tijssen, later published as a research article in a scholarly journal (Tijssen and Kraemer-Mbula 2018) and a policy brief (Tijssen and Kraemer-Mbula 2017); followed by a fulsome discussion with SGCs, which included experts Carlos Aguirre-Bastos from SENACYT (Panama) and Robert Felstead of UK Research and Innovation (UKRI).

This was followed by an international workshop that took place in Johannesburg in July 2018, supported by SGCI, and co-hosted by the University of Johannesburg and the Centre for Research on Evaluation, Science and Technology (CREST) at Stellenbosch University. The workshop deliberated on the experiences and reflections of scholars and practitioners from around the world, with a particular emphasis on those from, or working in, the Global South. Experts in attendance came from Asia, Latin America, Africa, Australia, Europe and the UK, and included representatives of funding organisations such as the NRF South Africa, NRF Kenya, Wellcome Trust (UK), UKRI and DFID, as well as key stakeholders such as the African Academy of Sciences (AAS) and some of their research partners across the continent. This workshop provided a fruitful platform to discuss early drafts of the chapters in this book, as well as collectively shape ideas for a future agenda of research excellence that includes the realities of the Global South. The meeting notably included several panels with invited researchers and funders operating across Africa, which infused our discussions with new perspectives and debates that significantly informed the chapters of this volume.

We wish to acknowledge the above organisations for their leadership, participation, support and insight during this event, with special thanks to the University of Johannesburg for hosting and supporting the organisation of the event (particularly to the Executive Dean of the College of Business and Economics, Prof. Daneel van Lill), as well as AAS for coordinating to have this event take place alongside the annual DELTAS meeting in the same location. We also wish to thank the following presenters and discussants, in addition to the contributors to this book, who were responsible for the rich feedback and discussions during these three days in July 2018: Dr Mark Claydon-Smith (UKRI), Dr Robert Felstead (UKRI), Allen Mukhwana (AAS), Dr Eunice Muthengi (DFID), Dr Simon Kay (Wellcome Trust), Dr Sam Kinyanjui (KEMRI), Tirop Kosgei (NRF, Kenya), Dr Glenda Kruss (HRSC), Prof. Rasigan Maharaj (Tshwane University of Technology), Prof. Johann Mouton (Stellenbosch University), Dorothy Ngila (NRF, South Africa), Dr Alphonsus Neba (AAS), Pfungwa Nyamukachi (*The Conversation Africa*), Dr Gansen Pillay (NRF, South Africa), Dr Justin Pulford (LSTM) and Prof. Nelson Sewankambo (Makerere University).

These efforts took place in parallel to the IDRC's dedicated work to advance how research for development is defined, monitored, managed and assessed. Many of these efforts have materialised in the Research Quality Plus (RQ+) approach as a tool that contextualises research quality and research evaluation for developing country contexts.

Overall, this book sets out to take a different approach from a standard collection of academic essays. It brings together people from a variety of settings and disciplines, and includes both practitioners and scholars. Many of the contributions are thus reflections on practical experiences, either from an individual or an organisational perspective. Editors and organisers of the 2018 workshop in Johannesburg from which most of the material is drawn sought to be 'reflexive' in the knowledge that is produced here. As we seek to broaden notions of scholarship, and argue for more pluralism, relevance and diversity, rather than decontextualised notions of excellence, we also apply this lens to our own work. We sought out outstanding contributions that bring new ideas that are relevant to the theme, but we chose not to 'standardise' the style or perspective taken by participants, preferring instead to have the contributions reflect discussions, debates and a collective search for solutions.

References

Tijssen R and Kraemer-Mbula E (2017) Perspectives on research excellence in the Global South: Assessment, monitoring and evaluation in developing country contexts. *SGCI Policy Brief* No. 1, December. https://sgciafrica.org/en-za/resources/Resources/SGCI%20Research%20Excellence%20Discussion%20Paper.pdf

Tijssen R and Kraemer-Mbula E (2018) Research excellence in Africa: Policies, perceptions, and performance. *Science and Public Policy* 45(3): 392–403

CHAPTER 1

Introduction

Erika Kraemer-Mbula, Robert Tijssen, Matthew L. Wallace and Robert McLean

Research excellence under scrutiny

Perceptions of what constitutes 'good science' shape the progress of knowledge creation and knowledge-based innovation. Globally, 'good science' affects decisions about what is funded, and what is not. It dictates who is rewarded and encouraged to pursue research. It promotes certain disciplinary traditions, but likewise discounts and discourages others. However, in the ever-competitive world of science and research, 'good' may not be good enough anymore. 'Excellent' science and associated prestige is increasingly seen as more valuable – something one should strive for. Not surprisingly, 'excellence' has become a buzzword, more popular than the underlying core notion of 'quality'. Those who are seen to be producing 'scientific excellence' are elevated to the highest paid jobs in the most prestigious institutions, granted greater degrees of academic leeway and expression, lauded as 'thought leaders' by peers, and turned to for policy and practice insights in the non-scientific realm. What gets called excellent, steers and influences the behaviour of individual researchers and teams, research organisations and research funders, and affects society at large. This would all be helpful and good if we had a widely endorsed view, and clearly measurable definition of, excellence. We do not. And it is highly unlikely we will be able to find a single suitable arrangement.

Nonetheless, there has been much high-level thrust for the adoption, application, implementation and celebration of 'research excellence' – at individual, institutional, and increasingly, national scales. In fact, excellence nowadays permeates all types of research and scientific work: from the curiosity-driven pure and discovery sciences, such as mathematics or logic to highly applied or translational work, such as epidemiology or anthropology. And the notion of excellence is permeating into research-related activities such as science communication, science-based education, knowledge translation and research management. What really makes for excellent science? How important is it we reach a consistent conceptualisation of excellence? Is excellence a means to 'protect' research against undue 'outside' interference, or a means of subjugating it to the requirements of managers, funders, publishers and other forces? And should striving for excellence be driven by the logics of competitive markets or by societal value considerations? These are important normative questions, and addressing them will require multiple voices, multiple perspectives and dynamic revisitation. This book attempts to add to this discussion.

There is a wealth of perspectives on excellence, and its implementation in science funding systems, that can be harnessed – from academics, non-academy-based scientists and non-scientists alike – to address those questions and feed this discussion. Take for example the adoption of the Research Excellence Framework (REF) in the United Kingdom, a high-income country with an advanced science system. This top-down REF approach provides performance-based funding to universities and promotes high-quality research through a quite explicit competitive scheme. It has gained considerable support from stakeholders in terms of increasing accountability and transparency, as well as promoting more rigorous standards. However, it has also sparked fierce criticism, especially from the UK's scientific community, for imposing an output-driven 'neoliberal agenda' and promoting over-competition within scientific disciplines that ultimately has an adverse effect on how contemporary science is produced, which is increasingly collaborative, interdisciplinary and impact-oriented.

Scholars from the humanities and social sciences have often been most vocal. These critiques touch on fundamental problems that extend far beyond the REF and the UK science system. Those from lower-income countries on the 'periphery' of world science also raise issues about their misrepresentation in scholarly journals and research disciplines, and the skewness of science in terms of its language and geographical distribution (see Vessuri et al. 2014; Chavarro et al. 2017).

Scientific research in lower-income countries, or in languages other than English, is poorly captured in most international databases and poorly covered by the main publishers who have come to dominate as gatekeepers and diffusers of research. These are some of the many biases that have become increasingly apparent. Cumulative advantage is another way that research from such countries or regions may be inadvertently considered less excellent, given how research resources are distributed globally, including both direct funding and access to infrastructure (equipment, library subscriptions, etc.). Scholars and scientists in lower-income countries also tend to face additional obstacles in their career development (lack of mobility, increased teaching loads) that restrict their ability to publish prolifically and to promote their publications.

The increased ubiquity of the term 'research excellence', its use in the context of rankings (at various levels), and the tendency towards quantitative scoring is not a coincidence. Nor is an increasingly explicit 'standardisation' of quality (e.g. through bibliometric statistics) at a global level, affecting most if not all disciplines and methodologies associated with scientific research. The standardised, global excellence paradigm makes it harder to play catch-up for given science systems, research-intensive universities, etc. that are relatively new, even if they are producing high-quality research. This move towards standardisation is problematic for assessing research produced in the Global South, in particular, as this is not where the standards originated. There is also evidence of a systematic bias towards researchers from the Global South in peer-review processes (see e.g. Yousefi-Nooraie et al. 2006). Clearly there is a need to deepen and enrich our understanding of excellence by presenting fresh views from academics and practitioners from the Global South, especially from those who have emerged relatively recently to take part in worldwide research structures, networks and disciplinary communities.

Being a common thread throughout this book, our use of the term 'Global South' requires some up-front clarification and explanation. Originating from the 1960s (Oglesby 1969), the term 'Global South' loosely refers to less developed or emerging countries. It is not meant to introduce a clear-cut dichotomy between the Southern and Northern hemisphere, nor between high-income countries and others in less developed stages of economic development. Our conceptualisation mixes both geographical dimensions and socio-economic characteristics. We use the term because it is a conveniently recognisable tag and a purposeful grouping of perspectives. When it comes to research excellence, the term represents a grouping that has been traditionally marginalised by more powerful voices.

In the remainder of this introductory chapter we set the stage by exploring some of the definitional issues around research excellence, and highlighting some debates and issues that have arisen in recent years, around the globe, related to the use of excellence as a normative term, the criteria used to judge it, and the far-reaching implications it may have. In essence, this book is an attempt to bring together critical voices from the often-overlooked science systems, particularly those of the Global South. We believe the reflections that follow will help to elucidate new debates and ideas on global and national scales, and that sharing and learning from these experiences and perspectives can bring about good change within the Global South, and around the world.

The elusive search for excellence

Using the term research excellence should, ideally, imply that it can be defined, recognised and assessed. Sometimes its meaning is obvious: for example, in describing important new discoveries or, on the other end of the spectrum, as a heuristic for sweeping narratives or impressive showcasing. However, more often, it escapes easy conceptualisation and identification. In everyday usage, the term excellence simply means being 'very good' (or at least 'better' than most others). Researchers who stand out above all others are seen as excellent. Focusing on excellence, as a normative concept, implicitly contains the assumption that it is possible to select the best proposals and best researchers by ranking. Excellence then implies determination by comparison, and therefore, competition (for research funding, for publications in top journals, etc.). Not surprisingly, excellence is often understood to be about elite science. Those 'best' researchers are not only masters of specialist fields, but are also creative and original. They are well positioned to determine what needs to be done in science and should be offered funding for their research proposals. Adopting such a narrow definition of the term also implies that it is possible to distinguish between a proposal for excellent research and one for non-excellent research.

Comparative judgements are of course unavoidable in circumstances where scarce resources are distributed and decisions require legitimation. Performance assessment is and will remain important, but we should strive to implement the best possible approaches. However, excellence is not a value-free term – far from it. It is highly contested and has acquired a set of specific meanings determined by dynamic interplays between science policy, funding instruments, research culture, performance assessment methodologies, internationalisation of science, and public accountability regimes. Building on the ideas of Gallie (1956), Ferretti et al. (2018) explore the idea of excellence as an 'essentially contested concept', highlighting the genuine difficulties that practitioners experience in coming up with a working definition for research excellence. In the extreme case, excellence could be construed as the degree to which a researcher measures up to his/her own values. Like the somewhat less problematic notion of 'quality', excellence is of course pluralistic and very much context sensitive. The evaluative criteria that make up quality in one field of scholarly work, (consider a pure math challenge that has stumped leading minds for decades) may not be the best criteria to judge research in another field (say clinical trials during a deadly disease outbreak). It is also time-dependent: what is considered 'excellent' today may well change dramatically in a few years' time. Accepting its inevitable fluid and multidimensional nature, there is still a need for systematic approaches to define and appreciate research excellence in order to manage science more effectively.

Some features of excellent science can be grasped and conveyed convincingly, and in many cases seem intuitive. Following the old truism 'what can be measured is treasured; what can't be scored is ignored', the quantitative approach tends to have more appeal and clout, especially among decision-makers craving clear and simple answers. In order to compare, performance must be observable and as measurable as possible. This urge for easily accessible information created a powerful drive for registering observable research outputs. Among the variety of approaches that have been used to identify and communicate research excellence during the last 30 years, the 'bibliometric' method has been particularly successful on a worldwide basis. Broadly speaking, bibliometrics comprises a number of quantitative analytic techniques that rest on the aggregation of quantitative 'indicators' captured from peer-review publication in journals indexed in international, largely privately owned, databases. A metrics-based approach requires yardsticks. Measuring the numbers of research publications in scholarly outlets, and/or the numbers of references ('citations') between publications, output levels were gradually adopted as a computational method to identify those top performers located at the high end of such performance distributions.

It was in the early 2000s that the citation impact approach was first explicitly connected to the notion of excellence, by assuming that excellence is more likely to be found in the top percentiles of citation impact distributions (Tijssen et al. 2002). Advances in bibliometric analysis methodologies, the increasing productivity of scientists (as measured by numbers of publications) and better ways of tracking these publications (e.g. through databases), since the first citation indices, have underpinned this particular attribution of excellence (or lack thereof). Nowadays, many bibliometric evaluation software tools, and also world university rankings, include a bibliometric indicator that refers to an entity's contribution to the 'top 10% most highly cited publications' as (an implicit) mark of outstanding performance. Supported by such (verifiable) empirical data, the empirical fact of being among the most highly cited worldwide can create an almost monolithic aura of exclusivity.

Many empirical studies have shown positive correlations between prolific output levels or high-impact performance and the outcome of *ex post* qualitative peer-review evaluations of scientific performance. However, questions about the validity and true meaning of bibliometric results, even when well executed, are coming to light too. These correlations often seem obvious, but it may prove difficult in some cases to disentangle cause (doing good research) and its effect (receiving citations as a mark of visibility, relevance or influence on others). For instance, the recognition from winning a Nobel prize 'causes' a significant number of citations. This is often referred to as a Halo effect or Matthew Effect, which refers to cumulative advantage processes that tend to favour those who are already prolific or highly visible in the international landscape of science. Citations alone can no longer be used as a predictor – other subjective factors prevail increasingly in the now exponentially large pool of 'top' researchers in a given discipline (Gingras and Wallace 2010).

Bibliometric approaches are valued for their (seemingly) precise results. And the straightforward quantitative ranking and comparison they facilitate is without doubt valuable for decision-making. But has simplicity seduced the system? Developed in the Global North, and based on a narrow concept of knowledge creation and sharing while extracting its empirical data from international sources that favour science in the advanced, higher-income countries, the 'top 10%' approach falls short in many ways. The citation impact approach provides at best interesting (but crude) comparative measures of excellence in 'discovery-oriented' science; that is, researchers working in worldwide communities on issues of widespread interest. It is certainly not very helpful for capturing scientific performance that addresses local issues or problems – be it applied, translational or discovery-oriented science.

The quest for excellence, rather than 'soundness' or 'quality', combined with the availability of quantitative indicators, often produces situations of 'hyper-competitivity' among researchers vying for finite resources and recognition. Such strong incentives to publish have been linked to the rise of predatory journals (which disproportionately affect researchers from the South), as well as increased cases of 'salami slicing' (publishing many separate articles instead of one of greater importance), 'ghost' authorship, and, in many cases, data manipulation and fraud. These trends, combined with evidence of lack of reproducibility of research in many fields and the exponential increase in publications, point to many incentives leading to greater *research waste* as well as the production of research which is less *relevant* to tackling urgent societal problems. Many have therefore urged the need to re-question and exercise restraint in the application of bibliometrics. Perhaps the foremost is the call to action for more responsible practice presented in the *Leiden Manifesto* (see Hicks et al. 2015, for the complete set of principles for action). Practical responses to the misuse of bibliometrics have also been launched; one leading example is the *San Francisco Declaration on Research Assessment* (DORA) which has recruited signatory members from across the globe to act out against bibliometric malpractice.

There is no international 'gold standard' metrics of excellence. Acknowledging the fact that it is definition-bound, assessment-specific and information-dependent, this book addresses a key measurement question: should research excellence solely reflect the criteria set by the scientific community, or should it reflect the broader value that we expect research to have for society? Opting for a broader and fluid concept of excellence requires developing measures able to capture multiple dimensions where we expect research to deliver social value. This process calls for joint efforts involving engagement and co-creation with relevant social actors. Such performance criteria also depend on geography – the location where the science is done, and where the primary users and potential beneficiaries of scientific findings are to be found. As one moves from a 'global' to a 'local' perspective, or from science in the Global North to that of the Global South, the core analytical principle should be: *scientific excellence cannot and should not be reduced to a single criterion, or to quantitative indicators only.* Any criterion of excellence in Global South science that does not take these considerations into account creates inadequate views and indicators of research performance, inappropriate assessment criteria, and therefore problematic rationales for justifying exclusivity of those tagged as 'excellent'.

Excellence becomes even more ambiguous when universities are described (or more often, self-described) as being 'excellent'. The above-mentioned REF, for example, or statistics on research publication performance, have shown an increasing focus on university rankings – and to a lesser degree country rankings – where the 'excellence' rhetoric hinders important debates and capacity building that should take place within these scholarly institutions (Moore et al. 2017). In the case of rankings, measurement of excellence is often done through a less-than-rigorous and often opaque methodology. Politics and public relations exercises blur debates on measurement methodologies. The question is often not 'how best to characterise the top universities' but rather, 'should we be ranking universities at all?'. And excellence does not necessarily only accrue to research outputs or impacts: high-quality features or outstanding performance may also emerge in knowledge sharing or dissemination strategies, ways of offering access to technical facilities, or other process-related characteristics of scientific research and its infrastructures.

University rankings are often prime instances of measurement out of context. Southern academic leaders have expressed concern that reliance on the predominant approaches to ranking may broadly miss the point for Southern institutions (Dias 2019). Worse still, rankings may exacerbate systemic bias toward the flawed approaches of the North, and undervalue unique ways of knowing, as well as essential scientific work from the South. Local relevance should be a leading concern and one of the key performance criteria, especially in resource-poor research environments of low-income countries of the Global South. A fuller picture can only be captured and revealed by applying assessment criteria and indicators that put researchers and users of research outcomes at centre stage. Adopting user-oriented approaches will require dedicated capabilities, cash and care. But it also needs a dose of creativity, and well-designed experimentation in the science funding models and mechanisms of the Global South is essential to arrive at workable assessment solutions customised for resource-constrained circumstances.

Indeed, the Global South may have a head start in developing and implementing these new and much-needed approaches. By avoiding the entrenched biases and well-described flaws of the mainstay methods of excellence assessment, Southern-derived solutions may offer potential improvements globally. One example is the *Research Quality Plus* (RQ+) approach developed by the International Development Research Centre (IDRC) with and for its Southern research community (Ofir et al. 2016; IDRC 2019). In short, RQ+ presents a values-based, context-sensitive, empirically driven and systematic approach to defining, managing, and evaluating research quality. As such, it is one practical and transferable response to the calls to action such as the *Leiden Manifesto* (see McLean and Sen 2019 for a comparison of RQ+ vis-a-vis the Manifesto's principles). But, as is argued within the dedicated chapter in this book (Chapter 15), RQ+ requires further trialing, testing and improvement. Still, the practical validation to date at IDRC, and at a growing number of Southern institutions, demonstrates that another way for research evaluation and governance is possible. A key purpose of this book is a further critique of, and experimentation with, new approaches such as RQ+.

Practical implications of embracing 'excellence' in the Global South

The Global South has an opportunity to do differently, and by doing so, to do better. Rethinking what makes for good science is essential; it is a process from which all can learn. But just as some of these issues can partly be traced back to the 'blind' quest for excellence, so too can new visions of excellence and quality have significant impacts on research systems, particularly in the Global South. In this book we present new options and alternative experiences. In the introduction outlined above we have only described the tip of the iceberg lurking beneath our collective scientific profession. It would be entirely possible for this book to focus solely on discontents with the status quo. But that is not our intent. Our goal is to provide a platform for new perspectives that have been under-represented and undervalued in the global debates and systems driving the status quo of excellence, and thereby offer novel experiences and different ways of thinking. We hope this lens will benefit those from either geographical location

(South or North), those across disciplines of science (pure maths or public health), or component (researcher, funder, university, government) of the global research system. We believe it opens a path toward a fairer, more efficient, more motivating, and more impactful global research ecosystem. In the following paragraphs we suggest why.

The adverse consequences of the quest for excellence are most strongly felt in the Global South, given scarce resources, and challenges in attaining visibility on a global scale. Moreover, the lesser developed regions of the globe also happen to be those where socially relevant research is most needed to address pressing local and regional development issues. Hence, more appropriate criteria and performance indicators, fit for purpose in the Global South, should embrace two other guiding principles: inclusivity and local relevance. As for inclusivity, with the rise of cooperation in science, and team-based research, it has become increasingly complex – and perhaps also less relevant – to assign a quality stamp to one particular 'excellent' entity, be it an individual researcher, an organisation or a country. Broader visions of local relevance can also help retain and reward a more diverse set of 'top' researchers, and thus a greater diversity of knowledge that can be assessed and compared. This can be achieved by recognising researchers' motivations for not only producing high-quality science (as judged by their international peers), but also pushing the boundaries of knowledge to tackle pressing societal problems (as judged by local society). To move in this direction, quality and excellence can be shaped to embrace a wider community of knowledge producers, brokers and users, reinforcing the 'social contract' that provides science with the autonomy and legitimacy to operate in the eyes of decision-makers, as well as the public. In an era where many point to declining trust in evidence and in scientists, this is sorely needed.

On a more practical level, accepting a pluralistic vision of research excellence can lead to greater flexibility in research evaluation practices and in setting research agendas that reflect development needs. This highlights the importance of science granting councils which, on a national scale, can link research to national policy priorities and facilitate connections between users and producers of scientific knowledge. This means putting the onus on useful, robust knowledge that can make a difference in a given context. While retaining what at times is a competitive process (e.g. to make funding decisions), research evaluation tools, particularly in the Global South, can be empowered to be more deliberate in recognising 'success' or 'quality'. Perhaps more importantly, moving away from a narrow or 'blind' usage of the term 'excellence' can enable funders to decide, based on evaluations as well as policy considerations, how to distribute research resources in a given system. In some cases, focusing on a few 'top' researchers or research teams may be desirable, while in others a greater return may be obtained from a more equitable distribution of resources (e.g. to promote diversity in approaches to solving grand challenges, or to build capacity in the research system).

What the South does not lack is scientific talent. Researcher capacity is another area where rethinking excellence, and how it is embedded in research systems, holds significant potential and importance for the future. However, few young people decide on a career in science in order to outperform other researchers in terms of the number of papers published or the popularity of their papers amongst other scientists. Instead, they develop an interest in scientific research – and make the difficult and at times costly choice to enter a career in research – motivated by a desire to do better for people, to advance a business objective, or even to benefit the health of our planet. But the academic incentive and rewards systems tend to favour, compensate and advance researchers based on the number of their publications, not on the socio-economic impacts of their research. This creates an often unnecessary tension between output-driven and impact-inspired science.

Of course, researchers will seek financial rewards for their investments and efforts, and feel good receiving the acknowledgement of their peers. But if these returns were tied to underpinning motivations (say to help people) rather than the insular status quo (such as the number of journal publications), a challenging and demanding career choice would receive renewed carrots for incentivising hard work. Measures of excellence which relate to the values and motivations of why people enter research would attract new entrants to research, and retain the fire and enthusiasm of those who do choose the path. On a global scale, there is a real opportunity here. As the world population grows it is expected that more than half of that growth will come from low- and middle-income countries. If Southern actors successfully align incentives to enter research with the right reasons for wanting to do research, there will be an unprecedented renaissance of science across the globe. At such a time, new ideas, advanced knowledge and fresh solutions will be most needed.

Structure of the book

Overall, this book sets out to take a different approach from a standard collection of academic essays. It brings together people from a variety of settings and disciplines, and includes both practitioners and scholars. Many of the contributions are thus reflections on practical experience, either from an individual or an organisational perspective. Editors and organisers of the 2018 workshop from which the material is drawn sought to be 'reflexive' in the knowledge that is produced here. As we seek to broaden notions of scholarship, and argue for more pluralism, relevance and diversity rather than decontextualised notions of excellence, we also apply this lens to our own work. We sought out contributions that bring new ideas that are relevant to the theme, but we chose not to 'standardise' the style or perspective taken by participants, preferring instead to have the contributions reflect a discussion, debate and collective search for solutions.

The volume thus seeks to address the needs of policy-makers, first among the granting agencies of sub-Saharan Africa, but also others around the world, to better grasp the issues, and to identify and implement policies and practices around research excellence to strengthen organisations and national research ecosystems. And as a result, the book should offer novel experiences and different ways of thinking that speak across geographies, disciplines and components of the global science system.

The first five chapters provide the theoretical underpinnings for new interpretations and uses of research excellence in the Global South. These contributions are critical to understanding precisely what the current problems are, what their current impact is on scholarship from the Global South and in identifying how rigorous, sustainable solutions can emerge and be implemented. Sutz sets the stage, calling for the need to move away from a 'universalistic' conceptualisation of research excellence that harms research agendas in the service of development objectives. Rather, she shows how alternative evaluation practices can better reflect these goals, in part by recognising excellence as 'situated' in specific institutions. Chataway and Daniels take stock of research-funding dynamics in Africa with a focus on science granting councils, and, taking into consideration the pressures faced by these councils, propose ways to 'embed' a new form of scientific excellence in the research they support, responding to a need for researchers' autonomy, while addressing national priorities. Tijssen's chapter draws on the body of knowledge that seeks to define and operationalise 'research excellence', highlighting new perspectives from the Global South that can lead to more nuanced interpretations of the term, as well as concrete recommendations for how research is evaluated. Kraemer-Mbula discusses the persistent gender disparities and imbalances in research performance, with particular attention to academic institutions in the Global South, proposing avenues to move towards diversity thinking in research excellence. Finally, Neylon portrays the current research excellence agenda as a manifestation of the dominance of international power centres at the expense of national or regional ties and information flows that are critical for development.

The second set of five chapters focuses on first-hand accounts of how universities, think tanks and granting councils currently operationalise the issue of research excellence. They shed light on the current constraints, trends and all-important national or regional contexts for implementation of policies and practices. The chapters highlight the need for grounding the conversation and for integrating new perspectives on the issue.

Siregar reviews the impacts and nature of policies of the Indonesian government to promote the quality and quantity of research in the country, pointing to a need to focus on research utility rather than a narrow view of excellence. Ouattara and Sangaré ground the notion of research excellence in terms of the formulation of research policies and instruments to promote high-quality, high-impact science in Côte d'Ivoire. Their experiences point to the need not only for effective processes in grants management, but also for broader efforts to strengthen national research systems. Ssembatya takes a longitudinal look at policies related to research excellence at Makerere University as the main research institution in Uganda, highlighting progress in many areas, but also policy gaps and perverse incentives that prevent the effective development of university research. Singh and Raza seek to explore new views of research excellence by examining some of the systemic biases that are faced by researchers in the Global South, bringing to the forefront different philosophies about research excellence, and finally arguing for the need to 'amplify' Southern voices when it comes to defining research excellence. Finally, Mendizabal provides an alternative view of research excellence through the lens of think tanks, which need to balance scholarly rigour and 'non-academic impact' in order to provide them with the credibility that they need to thrive.

The book's last four chapters – by Chavarro; Barrere; Allen and Marincola; and Lebel and McLean – focus on some of the tools and approaches that can be utilised to improve, or radically change, how research excellence or research quality can be interpreted and operationalised. This involves leapfrogging and leading the way from the Global South through innovative new platforms, policies and performance indicators. Through a re-examination of conventional research evaluation systems, Chavarro proposes putting 'sustainability' at the forefront of research evaluation systems, with a view to better tackling 'grand challenges'. Building on concrete examples of indicator development in Latin America, Barrere proposes broadening research excellence through the use of new assessment tools to measure the impact of research within and beyond the scientific community. Allen and Marincola focus on the scholarly publishing space as a means to offer powerful alternatives for research in the Global South to develop and utilise new tools to promote relevant and high-quality research. Finally, Lebel and McLean revisit the notion of research quality, using a flexible and holistic approach to assessing research for development, providing an alternative to 'conventional' views of research excellence.

A call to action, written by all contributors, concludes the book. It proposes a path forward, including how the term 'research excellence' should, and should not, be used, as well as how we might more broadly begin to develop and implement new ways of recognising high-quality, impactful scholarship from the Global South.

References

Chavarro D, Tang P and Ràfols I (2017) Why researchers publish in non-mainstream journals: Training, knowledge bridging, and gap filling. *Research Policy* 46: 1666–1680

Dias P (2019) Towards a ranking of Sri Lankan Universities. To rank or not to rank? *The Island*, 15 January. http://www.island.lk/index.php?page_cat=article-details&page=article-details&code_title=197761

Ferretti F, Pereira ÂG, Vértesy D and Hardeman S (2018) Research excellence indicators: Time to reimagine the 'making of'? *Science and Public Policy* 45(5): 731–741

Gallie WB (1956) Essentially contested concepts. *Proceedings of the Aristotelian Society* 56(1): 167–198

Gingras Y and Wallace M (2010) Why it has become more difficult to predict Nobel Prize winners: A bibliometric analysis of nominees and winners of the Chemistry and Physics Prizes (1901–2007). *Scientometrics* 82(2): 401–412

Hicks D, Wouters P, Waltman L, de Rijcke S and Rafols I (2015) Bibliometrics: The Leiden Manifesto for research metrics. *Nature* 520: 429–431

IDRC (2019) Research Quality Plus. https://www.idrc.ca/en/research-in-action/research-quality-plus

McLean RKD and Sen K (2019) Making a difference in the real world? A meta-analysis of the quality of use-oriented research using the Research Quality Plus approach. *Research Evaluation* 28(2): 123–135

Moore S, Neylon C, Eve MP, O'Donnell DP and Pattinson D (2017) 'Excellence R Us': University research and the fetishisation of excellence. *Palgrave Communications* 3 (January): 16105

Ofir Z, Schwandt T, Duggan C and McLean R (2016) *Research Quality Plus: A Holistic Approach to Evaluating Research*. Ottawa: IDRC

Oglesby C (1969) Vietnamism has failed ... The revolution can only be mauled, not defeated. *Commonwealth* 90

Tijssen R and Kraemer-Mbula E (2018) Research excellence in Africa: Policies, perceptions, and performance. *Science and Public Policy* 45(3): 392–403

Tijssen R, Visser M and Van Leeuwen T (2002) Benchmarking international scientific excellence: Are highly cited research papers an appropriate frame of reference? *Scientometrics* 54(3): 381–397

Vessuri H, Guédon J-C and Cetto AM (2014) Excellence or quality? Impact of the current competition regime on science and and scientific publishing in Latin America and its implications for development. *Current Sociology* 62(5): 647–665

Yousefi-Nooraie R, Shakiba B and Mortaz-Hejri S (2006) Country development and manuscript selection bias: A review of published studies. *BMC Medical Research Methodology* 6: 37

PART 1

Theoretical and conceptual underpinnings

CHAPTER 2

Redefining the concept of excellence in research with development in mind

Judith Sutz

The reasons behind the drive for excellence

In Mexico, at the beginning of the 1980s, a great devaluation of around 140% led to the plummeting of university employees' salaries, with the consequence, among others, of a significant brain drain. Raising salaries for all staff was not possible, and it was decided to give substantial bonuses to those considered more productive, giving birth to the Mexican National System of Researchers (NSR). Productivity was measured largely by publication count in and citations from ISI-listed journals (Neff 2018). An implicit concept of excellence was built. To be excellent in research for an individual researcher is to belong to the NSR, achieving the marks that the NSR considers proof of excellence. In the United Kingdom (UK), at the beginning of the 1990s, polytechnics were converted into universities. To avoid spreading resources over the whole university system, a competitive allocation for funds system was put in place and the weights used to measure performance were raised over time, to push further a process of differentiation (Cremonini et al. 2017). Again, a concept of excellence was implicitly built; it works exactly as the Mexican NSR works, defining who is excellent and why; that is, the place of excellence and how to get there. The irruption of the university rankings in the early 2000s unleashed what Hazelkorn

(2007) denominated a 'gladiator obsession' with the place occupied by national universities in the rankings. In Germany, following its poor performance in the 2003 Shanghai ranking, the Excellence Initiative was implemented, with the explicit goal of introducing further differentiation in the university system to achieve better research performance (Cremonini et al. 2017). In France a similar trend can be seen and for similar reasons, breaking a long tradition of equal funding treatment of universities through fostering a smaller group of universities 'that focus on excellence, have modernised governance, and are highly productive' (Hazelkorn and Ryan 2013: 90).

The current drive for excellence can be seen as a way, historically situated, to circumvent the limits that previous ways of assessing the value of academic work had for selecting fewer academics, academic departments, and universities. Becoming excellent has important economic consequences. Belonging to the Mexican NSR may imply a bonus of more than 50% of the total salary of a university professor. Being high in the Research Assessment Exercise (RAE), in the case of the UK, implied helping the university to rise in the rankings; this has immediate consequences in terms of the number of students, particularly foreign, coming to the university, whose fees cover around 50% of the university budget. These observations point to the need to consider the drive for excellence in context, the reasons why it appeared and some of the reasons why it endures. This helps to de-naturalise the drive for excellence, particularly in low- and medium-income countries (LMICs), as the right way to achieve capacities to create and use the best possible science for developmental goals.

The structuring effects of the strive for excellence

Excellence is a socially structured concept; it is also a socially structuring concept, once put into practice. Differentiation is at the heart of the social structuration of the concept; consequently, its structuring effects foster a race to not fall into the lower side of differentiation. A copious literature has analysed the consequences of this trend. '[I]nstitutions are measured against other institutions, researchers compete with one another for funds and universities for students. This leads to a permanent state of war between all the parties, destroying the social fabric of the university [...] Of all tasks in the academic workplace, teaching is the least appreciated and has to be outsourced as soon as possible, allowing people to focus on the battle for coveted research money' (Halffman and Radder 2015: 168). The striving for excellence in very different settings presents striking similarities in the effects of structuring. The Mexican NSR and the British RAE are good examples of that as both have been implemented for more than twenty years. In both cases, a 'unimodal' trend towards a specific type of research was found: results may be published in a given set of international journals strongly biased towards the English language. In both cases other academic functions were found to be given less attention, including teaching, institutional building and societal relationships (Foro Consultivo Científico y Tecnológico 2005; Martin and Whitley 2010).

The striving for excellence, even if its consequences appear similar everywhere, has become a dominant feature of science and university policies in the North and South for different reasons. Why bother with the place which 'Southern' universities achieve in the international rankings if they do not sell in the international market of higher education (HE) services? What is the use, in a relatively young, small and weak academic community, of signalling in different ways that only those who could be considered as scientists in the international community deserve to be considered scientists in the national community? There is an implicit argument behind these trends: Northern science (and its procedures) is a lighthouse, signaling the land in which Southern academics should try to arrive. These trends have been mercilessly described: '[T]he Third World looks to the North for validation of academic quality and respectability. For example, academics are expected to publish in Northern academic journals in their disciplines. Promotion often depends on such publication. Even where local scholarly publications exist, they are not respected. While it is understandable that small and relatively new academic systems may wish to have external validation of the work of their scholars and scientists, such reliance has implications for the professoriate' (Altbach 2003: 6).

A main point is that this type of mimetic behavior influences the science that is done and not done: Hess's (2007) concept of undone science is particularly relevant here. It seems fruitless to ask LMIC scientists to carry out the as yet undone science, relevant to their context, which nobody but they would attempt to work in, if the expected reward is lack of academic prestige and recognition, given that those interested in publishing the scientific results are mainly local or regional journals.

On the other hand, a main difference between North and South in this regard is the structure and dynamics of production. If imports – of artifacts or ideas – are the main and systematic way of solving problems in LMICs, the important legitimating source for research efforts implied in the expectation society has of its results is missing. The result is a push towards external approval, the trend described above. Lack of demand from the productive structure for indigenous capacities is one of the most serious sources of de-legitimisation of local science (and local innovation).

This problem was theorised more than 50 years ago by an Argentinean metallurgic engineer, Jorge Sabato. He proposed an 'interactionist' and systemic approach to the relations between science and technology and development, explained through a triangle (Sabato's triangle, widely used as a metaphor in Latin America), the vertices of which are Government, Knowledge Producers or Academia and Business Firms or Production. One of his main points is that more important than the strengths of individual vertices in relation to science and technology, the key for development is the strength of the interaction between them, the 'inter-relations'. Sabato also points out that each national system of science and technology is immersed in a wider international milieu; each vertex interacts with external actors through 'extra-relations'. When the inter-relations in a national triangle are weak, particularly affecting Academia, the concomitant isolation pushes the academic vertex to strengthen the extra-relations with the international academic milieu. Such extra-relations are deeply asymmetric: they are established between strong, well ingrained in society and legitimated science and technology vertices and those that are weak, isolated and barely legitimated. A vicious circle follows. The academic milieu of an underdeveloped country tends to adopt the agenda and academic legitimisation procedures of the highly industrialised countries, including predominantly their concept of 'research excellence'. This alienates even further their national integration; government and the productive sectors turn almost systemically towards foreign knowledge; the inter-relations within the triangle become even weaker; underdevelopment stays in place. Freeman used to call the trend of relying mostly on knowledge imports 'voluntary underdevelopment' (Freeman 1992). In Sabato and Botana's words:

> In a society where the triangle of relationships behaves well, the openings to abroad in the realm of exports of original science and technology or of adaptation of foreign technology produce real benefits in the short or in the long term.
>
> Historical experiences show that societies that have achieved the integration of the S&T triangle are able to produce answers and to be creative when facing external triangles of relationships.
>
> Very different is the situation, though, when the extra-relationships take place between dispersed vertices – not inter-related among them – and an external completely integrated S&T triangle. *This is one of the central problems that Latin American societies need to resolve, because in our continent [...] the base of the triangle shows an increasing and marked tendency to build independent relationships with the triangles of relationships of highly developed societies.* (Sabato and Botana, 1968: 23, emphasis added, author's translation)

Summing up: while the consequences of the prevailing striving for excellence are socially damaging in the North, they may be considered even more severe in the South.

A developmental view on research and excellence in research

As previously proposed, the concept of excellence in research is historically situated; moreover, it is ideologically moulded. In the case of universities, what counts as excellence in research depends on the aims

Figure 1: The asymmetrical relationships between academia in peripheral countries' systems and in highly industrialised countries' systems (base of the Sabato's Triangle conceptualisation)

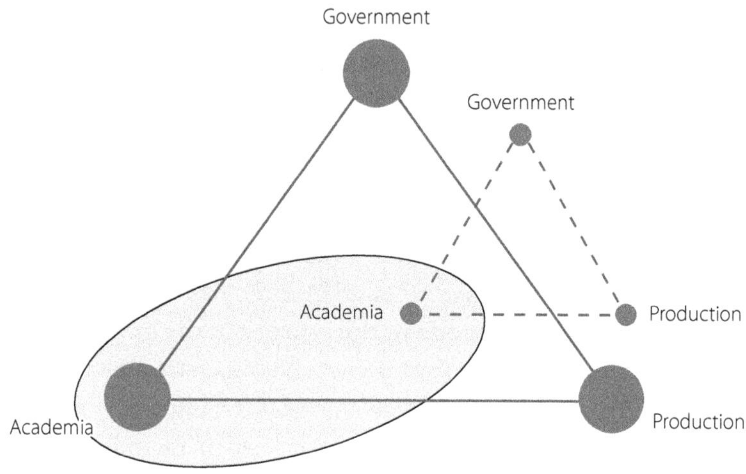

of the university. If the main aim were to climb the ladder in international rankings, the definition would be quite different from the one adopted if excellence were seen as maximising the impact of knowledge production on development. The latter has nothing to do with the often presented dichotomy between basic versus applied research; it relates to fostering a connection between universities and societal problems through the promotion of high-quality, relevant research and a tight relationship to high-level teaching and relationships with society. Developmental universities have been characterised in the following way:

> The '*Developmental University*' is characterised by its commitment to Human Sustainable Development by means of the interconnected practice of three missions, (i) teaching, (ii) research, and (iii) fostering the socially valuable use of knowledge. Such commitment means that developmental universities must contribute to building inclusive Learning

> and Innovation Systems by cooperating with other institutions and collective actors:
>
> (i) The teaching mission aims at generalising access to Higher Education, seen as lifelong advanced learning of increasing quality and increasingly connected with work, citizen activities, cultural expansion, and, in general, freedoms and capabilities for living lives that people value and have reason to value.
> (ii) The research mission aims at expanding endogenous capabilities for generating knowledge – at local, regional and national levels – in all disciplines and in interdisciplinary activities, with international quality and social vocation.
> (iii) The mission of fostering the socially valuable use of knowledge aims above all to cooperate with a wide variety of actors in interactive learning processes that upgrade the capabilities for producing goods and services as well as for solving problems, with priority given to the needs of the most deprived sectors.
>
> The definition could be given in a nutshell by saying that 'the Developmental University is characterised by its commitment to the democratisation of knowledge'. (Arocena et al. 2018: 169–170)

To the extent that the concept of excellence structures in part institutional aims, it seems clear that fostering developmental universities requires a specific conceptualisation of excellence. In particular, it can be said that more pluralism is needed to consider not only 'excellence in research' but also 'excellence in the search' of external actors with whom to build relationships conducive to a more useful utilisation of knowledge.

It is worth recalling that to serve developmental purposes research should be sound; mediocre results in scientific terms, regardless of the developmental importance of the topic, are useless. The soundness of a research effort and of its results should not be measured by proxies, such as the journal in which the results have been published or the scientific prestige of the proponents, even if these criteria may add arguments to a judgement based mainly on a direct appraisal of merits. A second assertion is that the questions and problems that research aims to solve are relevant criteria in judging how useful the results may be for development. This is not an exclusionary criterion: the need for science to answer fundamental questions within a discipline or to build a theoretical lens through which to better understand the world and the own reality is a legitimate goal for 'peripheral science'. This is a point worth stressing. Guillermo O'Donnell, an Argentinean political scientist, indicated that we should reject the pretension of some exponents of the dominant countries' academic milieu to consider that they speak from a sort of universal place, not recognising the particularities of other places by not recognising that they belong to a place too. He says, talking about Latin America but entailing a much broader scope:

> To conceive ourselves, in fact or right, as research assistants, as gatherers of data that are processed afterwards by theorists of the North, is equivalent to exporting raw materials with low value added to be processed by the industry of the North. On the other side, that of imports, this subordinate role means to 'apply' mechanically theories already developed in the North, which is equivalent to importing turn-key industries or technologies to which at most some adaptations are made. (O'Donnell 2004: 8, author's translation)

From a developmental perspective, excellence in research needs to be considered from a different angle than the one analysed so far. Of course, we may dispense with the concept of excellence, given the meaning it has acquired, using instead 'quality research', for instance. A recent work analyses research excellence as a 'contested concept', showing unmistakably the inherent complexity involved in its characterisation (Ferretti et al. 2018). The term we use is not however the important thing. The question is through which attributes do we spot those research projects, research programmes or individual researchers that deserve support from a developmental point of view? After that, we must consider the question of how to proceed to select among them the few that will receive support. First, those considered to be excellent or of high quality should be identified.

We may have 'relevant attributes' and 'not so relevant attributes' in assessing research proposals from a developmental perspective in LMICs. For instance, aiming to be published in *Nature* or *Science* and presenting a programme to achieve that aim is not a relevant attribute; strengthening the physics community – theoretical and experimental – through building research groups devoted to some of the fundamental branches of the discipline in a country with very low capabilities in the field is a relevant attribute. The dichotomy between 'the best and the rest', implying that the rest is worthless from a scientific point of view, is not acceptable.

The 'teaching trickle-down' effect of a research proposal or of a researcher's activity is a relevant attribute. It can be indirect, by strengthening a weak research area, thereby allowing senior researchers to teach creatively and raise creativity among their students; it may be direct, by adding new perspectives to a current course or even by developing new courses. The importance for concrete stakeholders of the problems addressed is also a relevant attribute. Originality is an important attribute; sometimes the value of a proposal from a developmental point of view is the degree of deviation from orthodox approaches. The number of young people substantially involved in a research proposal is a relevant attribute, as is non-subordinate participation in international networks.

There is not a single set of relevant attributes, valid in all circumstances, even though the few just mentioned may be considered useful in general. Countries have different needs in terms of the knowledge required to advance developmental goals and relevant attributes should take this into account. This also applies to the strengths of the research community, which may put a premium in certain directions if they promise to start redressing important weaknesses.

A funding agency needs clear assessment criteria to be fair and accountable. To combine this with 'developmental soundness', the basket of relevant attributes at its disposal should be sufficiently ample and well fitted to the unit of analysis. Building such a basket is a fine work to be carried out by funding agencies, in cooperation with the beneficiaries to devise the attributes that proponents should highlight in their proposals. This points to a situated redefinition of excellence in research, taking developmental goals into account (Arocena et al. 2019).

A weak scientific community in a small peripheral country with an unsatisfactory innovation system: How to do good through research policy

Uruguay is a high-income country according to the World Bank classification, based on per capita income. Other indicators are as follows: research and development (R&D) GDP is 0.35; participation of development in total R&D efforts (the other two components being basic science and applied science) is 13%; participation of business firms in R&D investment is less than 30% (including public firms in the oil, electricity and telecommunications sectors); researchers working in business firms are fewer than 5%; the number of researchers per million inhabitants is slightly over 500. A rapid comparison with other small high-income European countries shows important differences in all science, technology and innovation (ST&I) indicators; the other Latin American country in this league, Chile, shows the same ST&I figures as Uruguay. Clearly, high per capita income is not necessarily a good predictor of good science and technology (S&T) activities; the other way around makes more sense empirically.

All LMICs show poor performance in S&T indicators. Some of them are extremely poor; other are not so poor but are extremely unequal (e.g. many Latin American countries); in general, their endogenous efforts towards enhancing S&T capabilities are low. Even when efforts are made to increase HE enrolment, there are no concomitant efforts to find productive and creative jobs for graduates. Usually, the most complex and intellectually challenging problems are solved via imports or consultancies from abroad; the long and expensive process of building local capabilities to solve problems is thus weakened. Moreover, the configuration of innovation systems in LMICs shows weak interactions among actors and missing actors as well.

The question about how to 'do good' through research policy in contexts such as those described above cannot be answered by a cut and paste from recommendations prepared for other realities (as is often the case). Diversity conspires against general principles, but some can be proposed.

First, the whole gamut of the national research community needs to be strengthened. This is fundamental to achieve a healthy research ecosystem. However, there is no single instrument to do this, because in any quality-based competition for funds it will not be possible to avoid the 'Matthew Effect', particularly so when strong asymmetries among fields of knowledge, research groups and individual researchers are present. Specific programmes to enhance the quality of research in weak fields of research are important. They need to plan in the medium term, be based on sound appraisals of the current situation, emphasise raising the academic level of researchers, and be monitored continuously to detect problems early.

Second, international exposure needs to be enhanced, although not only by sending local people abroad. A dynamic of local seminars, workshops and conferences with the participation of invited professors from abroad may be more 'spreadable' in terms of benefits for the national research community.

Third, demonstration effects are important in places where local capacities for knowledge production and problem-solving are not much valued. Low morale is a problem for researchers in LMICs; the belief that only by being praised abroad can they be recognised as good researchers is an obstacle to reconciling research excellence and developmental goals. Reversing self-defeating imaginaries in relation to S&T is a very difficult cultural challenge in which several actors need to be involved. Interdisciplinary research teams convoked to work on problems where their contribution may make a difference can help to give visibility to research as a problem-solving tool and to local researchers as problem-solvers.

Some general working principles developed at the University of the Republic's Research Council

The University of the Republic was until some years ago the only public university in Uruguay; it is the only one which cultivates all research fields and grants professional education in all fields of study. In terms of research, combining all current indicators, it is responsible for around 75% of the academic knowledge produced in the country. The University of the Republic is an uncommon institution, sharing only with Argentinean public universities its identity features: it is free of charge; all those who finish high school are entitled to enter university, regardless of their past academic performance; and they may choose freely in which faculty they want to study, without any limitations (*no numerus clausus*). There are other academic institutions devoted to research, but they are concentrated in the life sciences.

The military dictatorship (1973–1984) included the military rule of the university and the destruction of almost all the national academic fabric; the migration rate of the academic staff during these years was huge.

In 1992, the University Research Council was created; it was endowed with a budget with the mandate to help reconstruct and enhance university research. It is a 'central' body of the university governance structure, meaning that it is, in principle, independent of the will and policies of individual faculties. The council operates mainly via competitive calls for academic activities related to research.

The evolution of the academic fields since the reconstruction of the university's autonomy that accompanied the recovery of democracy was very uneven. Exact and natural sciences were able to recover and grow quite rapidly; clinical research was much more difficult to strengthen; agrarian sciences and technologies had mixed outcomes, as did social sciences and the humanities. Within each field, disparities were also significant. So, both a goal and a foe were identified. The goal was to strengthen research capacities in all fields and subfields; the foe was the Matthew Effect that lies in wait to concentrate resources in those better-off disciplines if attention is not paid to its dangers. The way to achieve this emerged from a consensual common sense built over time within the Research Council and, more importantly, within the evaluation committees convened to work on the appraisal of the proposals presented at the Research Council's different calls. This common sense can be summarised as follows: to allow research evaluation to make room simultaneously alongside academic quality and research policy goals. This entails a compromise, particularly on the side of research evaluation, implying that not necessarily the best – designed as such by an agreed mechanism – will necessarily be those chosen for support. This is formally recognised in the texts of the Research Council calls: 'efforts will be made to assure that all disciplines and subdisciplines are represented in the results of this call'.

The mechanism to achieve this was to visualise a 'band of acceptable research quality' outside of which proposals are rejected due to lack of merit and within which proposals of relatively similar merit are considered. This implies that if proposal x in discipline A, which for the first time would receive support to perform research activities, falls within the band, it may be given precedence over proposal y in discipline B, which has several good proposals, even if the evaluation received by proposal x is not as good as that received by proposal y.

This mechanism helps to avoid the Matthew Effect. Another procedure with the same aim is to try to establish competence between proposals and not between proponents. The CVs of the proponents are used mainly to ensure that there is sufficient scientific capacity to lead the research to harbour. Neither of these mechanisms is easy to implement, and in each evaluation round it must be remembered that they are 'official policy'. However, over time a shared evaluation culture takes precedence over simply picking the best proposals, leaving the Matthew Effect to operate freely to the eventual detriment of younger researchers and less well-developed areas of research.

Another policy guide for the Research Council is that there is no single research policy instrument, regardless of how well conceived, which is able to address the diversity of policy aims. In a weak scientific community, it is probable that whole fields of knowledge or disciplines or sub-disciplines fall outside the 'band of acceptable research quality'; this is certainly the case in Uruguay. These will continue to fall outside this band unless specific measures are taken to allow them to improve their research capacities, as a healthy research ecosystem requires.

A programme aimed at this type of goal has already been mentioned. In the Uruguayan case, a programme called the 'Enhancement of the Quality of Research in the Whole University' was put in place. It starts with a self-appraisal of research weaknesses with the support of a foreign expert; then a four-year 'enhancement of the quality of research plan' is elaborated, establishing annual goals; finally, the deployment of the approved plans is accompanied by a special group of researchers, who monitor the advances and detect early problems. The 'units' of this instrument may be whole fields of knowledge, such as psychology or weak parts of a strong field, such as the medical-physics field. This is an expensive instrument; it directs important resources to the weakest part of the university's research capacities amidst budgetary constraints. Nevertheless, it has won legitimacy at the university as a whole because there is a consensus that research weaknesses that need to be redressed can be found everywhere.

Finally, two additional guiding ideas for the Research Council are that early career researchers and 'the best' need specific support. Regarding the latter, it is worth stressing that avoiding the Matthew Effect should not imply 'leveling down'. Those areas of research that excel need to be supported by giving them breath to work over medium-term programmes; this is done by means of a four-year funding scheme directed to consolidated research groups. Support for early career researchers has proven to be a tricky issue, because what is considered 'early career' varies among cognitive areas and institutional trajectories. In fact, along the fourteen editions of the programme devoted to young researchers, the definition of the target kept on changing, according to a better comprehension of what 'young researcher' means, and due to institutional changes that affect that meaning.

'Plural evaluation/engaged evaluation' or how to assess proposals oriented to developmental goals

Managing the programme 'Research and Innovation Oriented Toward Social Inclusion' is quite difficult for the University Research Council.

The difficulties stem from various sources, of which the evaluation process is not the smallest. First, there is a need to assess the degree of social engagement of the research proposal; that is, to what extent the research tackles a problem of social exclusion recognised as such by some involved stakeholders. This provides key information to evaluate whether the proposal has merit to belong to the programme. If the research problem appears to be of interest mainly for the research team, then the proposal is rejected before any academic appraisal. The information is gathered through personal interviews with the stakeholders indicated in the proposals. Sometimes the interested stakeholder has the power to incorporate the research results into its practices, typically when public policy is involved. Other situations require mediations to put results into practice, in which case mediators are interviewed to assess, first, if they have been contacted, and second, to what extent they are willing to assure the needed actions to implement the research results. Once this 'engaged part' of the evaluation is completed satisfactorily – that is, it is confirmed that the research proposal tackles a problem that is considered as socially exclusionary by a concerned stakeholder and that the actors who may facilitate the application of the results have confirmed their engagement – the proposal passes to 'ordinary' research evaluation. The academic merit of the proposal is appraised through the justified opinions of two reviewers, generally foreign, given the small size of the local research community. Once at this stage, the process regains its classical form, with academic quality measured through the usual indicators defining the evaluation outcome.

The combination of these sources of information helps to spot loopholes in the proposals that may then be discussed with the proponents, should the overall merit of the projects suggest the convenience of supporting them. The proposals presented to this programme are much more difficult to prepare than ordinary R&D projects and so the volume of demand is low. The social commitment of the university explains the efforts made not to lose a good project if it could be reasonably reformulated.

This programme aims, of course, to help social inclusion with the concourse of research. But more fundamentally it aims at helping researchers to become aware of and interested in putting their knowledge at the service of social inclusion. At some point, it was understood that researchers frequently needed to reflect thoroughly about a series of matters before being able to prepare a proposal. They needed, for instance, a better knowledge of the perspective of stakeholders in relation to the way they were seeing the problem; sometimes they needed to make sure that the methodology through which they wanted to tackle the problem was accurate enough. So, a second entry point to the programme was put in place, namely, the presentation of a short proposal to explore and clarify the aspects needed to prepare a fully fledged project. The evaluation of this modality also follows a plural path: first, the evaluation committee assesses the social merit of the proposal and then experts are required to evaluate its scientific quality.

These 'plural' and 'engaged' evaluation processes are extremely time consuming and can be implemented if the number of proposals is small. However, the experience gathered from them feeds reflexive appraisals of the dynamic of research that help to refine research policy instruments aimed at developmental goals.

An ongoing struggle and a needed redefinition of excellence

Turning now to individual researchers, an NSR was implemented in Uruguay in 2008, providing a 'categorisation by excellence', accompanied by a monetary reward according to the category achieved. At the university level, where the vast majority of researchers work, a 60-year-old stimulus regime grants a 60% rise in salary to those devoted full time to university activities – including undergraduate teaching – with particular emphasis on research. The conflicts between the evaluation criteria of the NSR and those of the university regime became rapidly apparent. Not only does the NSR concentrate exclusively on research and postgraduate teaching, but its main criteria to appraise research activities relate to the number of publications in international journals, or international editing houses, in the case of books. The evaluation relies on the information provided by a normalised CV form. On the one hand, to climb the hierarchy of the system – and avoid exclusion from it – it is fundamental to gain international visibility through publications in recognised journals or through high citation counts. On the other hand, even if research is particularly important in order to gain grants for the university, it is not the only activity that counts. Moreover, the diverse traditions of knowledge production and communication within the university are recognised, and so plural evaluation criteria are put in place, including the direct appraisal of a piece of work selected by the applicants, besides the information included in activity reports and CVs.

Around 80% of all full-time university researchers also belong to the NSRs. Even if in economic terms full-time work is significantly more important than the NSR, the latter started 'colonising' the evaluation criteria of the former. Part of this stems from the 'external' character of the NSR, supposedly less affected by inbreeding than the university regime. However, in a small academic community, where the evaluation committees of the NSR consist almost exclusively of university researchers, this argument is more rhetorical than real. But perhaps more important is the idea that the NSR spots the best, while the university full-time regime supports researchers who perform well and with high intensity, but do not necessarily strive to belong to any ranking. Attribution of academic prestige within the country according to how near researchers are to be considered excellent by international standards has proved to become, in a short period, the most powerful tool to discipline researchers into the NSR path, particularly the younger ones.

The 'regime of prestige' of the NSR overpowered that of working full time in the university, which used to be highly valued. The problem is, as in so many other similar experiences, that university activities such as teaching, which take time from research, began to be seen as burdensome if mandatory, and were simply left behind if they were voluntary, such as institutional building or community service. To countervail this trend, it was proposed in 2012 to give full-time researchers in the university the freedom to choose plural research paths. They may tackle complex problems without accumulating publishable results in the evaluation period and nevertheless be highly regarded, if their working strategies are sound. They may produce one good paper and devote the rest of their time to performing meaningful and difficult tasks, such as preparing a new masters programme or building relationships with external actors to be able to address some of their problems. In short, a signal was given that the university considers highly valuable the fact that its researchers combine quality research with quality performance of other academic and social activities, based on their research capacities.

The proposal, even though formally approved, encountered fierce opposition from influential researchers, with the argument that its application would undermine the quantity and quality of university research. The idea that the quantity of papers in international journals should not be a main evaluation criterion was particularly contested. Nevertheless, uneasiness started growing from below as time went by. Some senior researchers were surprised by the reluctance of their students to tackle complex problems in their PhD theses, giving the argument that they needed to publish quickly; others recognised increasing academic misbehavior associated with 'salami' papers, co-authorship cooperatives and so on. For researchers in some disciplinary orientations, the tension between the NSR requisites and their vocation to tackle problems of national importance became a real problem.

Discussions around the evaluation of researchers on how to appraise excellence, taking into account the national context, or on how to reconcile quality research with the aim of achieving developmental goals, have gained momentum. The growing international criticism of the prevailing research evaluation practices helps to put aside dismissive arguments against those who locally criticise such practices. Pluralism seems to be slowly recognised again as an important feature of a research evaluation system that makes room for diversity, for interdisciplinarity and for social engagement. In a recent workshop on the subject, organised by the University Research Council and attended by an important number of researchers, a message that resonated with force and was taken up by many was 'one size does not fit all'.

It is interesting to note that the conflicts around research policy are not centered on policy instruments: for instance, programmes devoted to social inclusion or to the public understanding of problems of general interest in society are not accused of deviating scarce resources from the pressing needs of excellent research groups. The conflicts are centered on how to appraise individual merits, and on how to give and earn academic prestige. How this conflict is resolved has consequences on the demands made on research policy instruments: those instruments that allow a focus on the type of academic work that is praised by the individual research evaluation criteria will be overselected.

There is a complex web of interactions between research policy instruments, the evaluation criteria of individual researchers, the decision-making of a single academic unit taking these two dimensions into account – for instance a university – and decision-making at supra-levels which have their own criteria, national or international. This complex web of interactions does not work smoothly towards a common end. The Matthew Effect, for instance, is something that can be detected at local level; it is more difficult to perceive it at national or international levels. As already mentioned, national criteria, which strive to achieve international visibility for national science, may jeopardise efforts made at local level to better produce knowledge related to developmental goals.

Achieving a minimum level of consensus around a redefined meaning of research excellence – a counter-hegemonic meaning – is important to avoid weakening, by the overpowering of some meanings over others, the directionality of research policies aiming at developmental goals. This is an extremely complicated task involving ideological aspects, as well as more technical ones. Telling a developing country that trying to play in the great leagues is not a reasonable goal may be seen as a recommendation which has a colonial mindset; a much more productive approach would be to legitimise the variety of small roads by which science may contribute to human well-being.

A mutual comprehension of the problems involved in any redefinition of research excellence needs dialogues among the different stakeholders of research policy, international, national and local. In some countries, interesting exercises of research evaluation involving academics and non-academics have recently taken place. Something similar could be done in Latin American countries, as an experiment, allowing actors to work together across these different research policy levels. This striving for plurality in research evaluation would imply, in present times, sailing against the strong wind of quantified homogenisation, and would unite concerned researchers in the North and South, which holds promise of change.

References

Altbach P (ed.) (2003) *The Decline of the Guru. The Academic Profession in the Third World.* New York: Palgrave Macmillan

Arocena R, Goransson B and Sutz J (2019) Towards making more compatible research evaluation with developmental goals. *Science and Public Policy* 46(2): 210–218

Arocena R, Göransson B and Sutz J (2018) *Developmental Universities in Inclusive Innovation Systems. Knowledge Democratization in the Global South.* London: Palgrave McMillan

Cremonini L, Horlings E and Hessels L (2017) Different recipes for the same dish: Comparing policies for scientific excellence across different countries. *Science and Public Policy* 45(2): 232–245

Ferretti F, Guimarães AP, Vértesy D and Hardeman S (2018) Research excellence indicators: Time to reimagine the 'making of'? *Science and Public Policy*: online. https://doi.org/10.1093/scipol/scy007

Foro Consultivo Científico Tecnológico y Academia Nacional Mexicana de Ciencias (2005) *Una Reflexión sobre el Sistema Nacional de Investigadores a 20 Años de su Creación*

Freeman C (1992) Science and economy at the national level. In: C Freeman (ed.) *The Economics of Hope*. London: Pinter. pp. 31–49

Halffman W and Radder H (2015) The academic manifesto: From an occupied to a public University. *Minerva* 53(2): 165–187

Hazelkorn E (2007) *How Do Rankings Impact on Higher Education?* OECD Programme on Institutional Management in Higher Education

Hazelkorn E and Ryan M (2013) The impact of university rankings on higher education policy in Europe: A challenge to perceived wisdom and a stimulus for change. In: P. Zgaga, U Teichler and J Brennan (eds) *The Globalization Challenge for European Higher Education: Convergence and Diversity, Centres and Peripheries.* Frankfurt: Peter Lang. pp. 79–100

Hess D (2007) *Alternative Pathways in Science and Industry. Activism, Innovation, and the Environment in an Era of Globalization.* Cambridge, Massachusetts: The MIT Press

Martin B and Whitley R (2010) The UK research assessment exercise: A case of regulatory capture? In: R Whitley, J Gläser and L Engwall (eds) *Reconfiguring Knowledge Production: Changing Authority Relationships in the Sciences and their Consequences for Intellectual Innovation.* New York: Oxford University Press. pp. 51–79

Neff M (2018) Publication incentives undermine the utility of science: Ecological research in Mexico. *Science and Public Policy* 45(2): 191–201

O'Donnell G (2004) Ciencias sociales en América Latina. Mirando hacia el pasado y atisbando el futuro. *LASA Forum*, vol. XXXIV, no. 1. pp. 8–13

Sabato J and Botana N (1968) La ciencia y la tecnología en el desarrollo futuro de América Latina. *Revista de la Integración* 3. Buenos Aires. pp. 15–36

CHAPTER 3

The Republic of Science meets the Republics of Somewhere: Embedding scientific excellence in sub-Saharan Africa

Joanna Chataway and Chux Daniels

Introduction

Research excellence is often equated with publication in journals which have a high-impact factor. Yet ample evidence exists of distortions associated with defining research excellence solely in relation to publishing breakthrough research in high-impact journals. A recent study, conducted in the context of the African Science Granting Councils Initiative (SGCI), reviews the issue of research excellence in sub-Saharan Africa (SSA) and the need for an approach which expands the notion of excellence beyond publications altogether (Tijssen and Kraemer-Mbula 2018). The SGCI is a multi-funder initiative that aims to strengthen the capacities of science granting councils (SGCs) in SSA in the management of research, the design and monitoring of research programmes, knowledge exchange with the private sector, and partnerships between SGCs and other science system actors. SGCs[1] refer to science councils, research councils or agencies responsible for the funding and/or management of science and research in SSA.

The study by Tijssen and Kraemer-Mbula (2018) reveals that publications in high-impact-factor and influential journals are thought by many SSA actors to be important. However, in relation to defining excellence in research, other factors were judged to be equally important. These factors include potentials for, or the ability to, generate significant societal impact, research relevance or research alignment with socio-economic objectives, the choice of indicators (or metrics) and the research criteria being evaluated.

A clear challenge is the need to construct measures of performance and evaluation which foster research that relates to social, economic and environmental challenges. Such measures of performance and evaluation must be aligned to national-level SGC attempts to build capacities and capabilities (AOSTI 2013; Chataway et al. 2017a), and knowledge (AAS 2018) that align with SGC missions [2] to contribute to national development agendas and science, technology and innovation (ST&I) policies in SSA (AUC 2014, 2015). At the same time, the Tijssen and Kraemer-Mbula (2018) study highlights a clear desire by researchers and funders to promote the production of rigorous and high-quality research.

The discussion about whether, given this complexity, conventional metrics (e.g. number of publications and ranking, or citations) should be used as the sole criterion for research evaluation is closely aligned to a broader discussion of whether academic peer review is an effective mechanism with which to judge academic research. Although metrics is often correlated with peer review, the two issues, although sometimes conflated, are not the same. They can also have quite different implications. One approach has been to treat them as a sort of trade-off between the autonomy and strength of the academic community. In this trade-off approach, the strength of the academic community, often operating at an international level, is at odds with the power of other actors, often local, to get their voices heard in relation to the quality and relevance of knowledge production. The two sides of the argument are referred to in the title of this chapter as the 'Republic of Science' and the 'Republics of Somewhere'.[3]

In this chapter, we explore the idea that the discussion does not necessarily have to hinge on that classical trade-off approach and narrative. Although more work needs to be done, the work of Tijssen and Kraemer-Mbula begins to demonstrate that often researchers and funders want to reconcile 'excellence' and 'relevance'. The underlying tension then looks different. When the academic community and the SGCs that support them have insufficient autonomy and 'capital' in their national environments, they are limited in their capacity to embed their research effectively in addressing societal challenges. Looked at from this perspective, the issue of autonomy is related to the variety and strengths of 'capitals' and capabilities that SGCs, and the researchers which they support, can draw on in their role as national actors.

The Republic of Science: Autonomy and peer review

The following section of the chapter links debates around scientific autonomy and embeddedness or relevance to challenges facing SGCs.[4] As background to that section, it is useful to briefly reflect on publishing, peer review and definitions of 'excellence' (Benner 2011). The Republic of Science is a fascinating and powerfully argued essay authored in 1962 by Michael Polanyi. In the essay, Polanyi sets out arguments in favour of high degrees of autonomy and freedom in relation to governance structures for scientists and science-funding bodies. The influence of The Republic of Science notion of scientific excellence continues to influence modern debates in science and research. Under this notion of excellence, academic peer review is a key mechanism through which academic autonomy is exercised.

With respect to dealing with the undue influence of metrics and impact factors and the need to open up publishing options, the pressure to reform could be seen as one of reform of the Republic by its own citizens. In this formulation, academic peer review is retained as a key role and this ensures high degrees of autonomy. From this perspective, the Republic has become corrupted in a sense by the power exercised by particular publishing regimes and conventions. Reform does not necessarily signify revolution in relation to governance of the Republic and academic peer review can still be viewed as the bedrock for excellence, but within the context of a changed approach to the importance attached to impact factors. Many open science initiatives such as those hosted by F1000 and the African Academy of Sciences are examples of this reforming approach.

So, from this view, after reform of a publishing system gone awry, the autonomy of scientists to determine what is excellent can remain more or less intact. However, in this chapter, we focus on the related but different problems and tensions which arise in relation to securing mechanisms to ensure relevance and embed research excellence in national contexts, while protecting the autonomy of scientists. It is useful to separate out these two issues because with respect to increasing the immediate relevance of science, more radical reform of the Republic might be needed with 'non-scientists'; that is, non-academic, taking a greater role in the determination of excellence. For many scientists this is more challenging and in extreme forms can undermine the authority and autonomy of scientists. The following part of the chapter looks briefly at some of these debates and lays out particular ways in which The Republic of Science is challenged by national agendas or The Republics of Somewhere.[5]

The concluding part of the chapter develops some preliminary thinking about how research councils – namely SGCs, in the context of this chapter – can orient themselves in the context of needing to respond to the critiques of conventional assessment and its foundations, which are related to the 'Republic of Science' model of research. We outline some thinking, which underpins a notion of embedded excellence as an alternative to the notion of excellence based on publications, or on the distinction between applied and basic research. We suggest some practical ways in which that concept might guide the work of SGCs in SSA, but also of science councils elsewhere

SGCs: Between the Republic of Science and the Republics of Somewhere

An implication of the opening paragraphs of this chapter is that we might relate debates about the tensions between scientific autonomy on the one hand and relevance and embedded excellence on the other hand, in part at least, as an issue of national versus global and regional level decision-making authority. Viewed from this perspective, SGCs have a key role to play in resolving and negotiating different demands made on science/research and researchers. This section explores the role of SGCs in more detail.

To reiterate, science granting councils (SGCs), as used in this chapter, refer to organisations that fund, direct or manage science and/or research in 15 countries in SSA. These countries are part of the Science Granting Councils Initiative (SGCI) set up and funded by Canada's International Development Research Centre (IDRC), the United Kingdom's (UK) Department for International Development (DFID) and South Africa's National Research Foundation (NRF) (Chataway et al. 2017a, 2019). The objective of the SGCI is to strengthen the SGCs' ability to manage, design and monitor research programmes; promote and support knowledge exchange with key ST&I stakeholders; and establish and foster partnerships among SGCs and ST&I stakeholders.

In order to carry out these activities, SGCs need to utilise robust ST&I indicators and metrics, and engage with ST&I ecosystem actors, comprising the private sector, funders, policy-makers and scientists or researchers. The need to engage with a wide range of actors highlights the issue of retaining autonomy for scientists, while relating to national policy agendas and national priorities. As mentioned earlier, Polanyi's 'The Republic of Science' is an impassioned plea for scientists to be given the freedom to determine research agendas and to judge scientific excellence (Polanyi 1962; Rip 1994; Flink and Kaldewey 2018; see also Bush 1945; Benner 2011). Over the decades, these ideas have been called into question from a number of angles and perspectives. These critiques point to the flaws in the classic 'autonomy framing' and the priority it gives to academic peer review. They also highlight the flaws in associated 'linear model' thinking. In relation to these arguments, various schools of thought associated with the nature of innovation systems and socio-technical systems have emerged. A recent debate in *The New Atlantis* provides powerful arguments against some of the fundamental constructs of the Republic of Science (Sarewitz 2017[6]) and, on the other hand, concern that the approach ignores the importance of serendipity in scientific findings and research (Curry 2017).

A group of research and innovation scholars have pointed to the gains for researchers and research funders that can come from defining themselves in relation to social contexts in which they exist.

These researchers and research funders can, in addition, promote overall visions for national and global sustainable development agendas that are more inclusive and do not exacerbate challenges such as inequality and environmental degradation (de Saille 2015; Arocena et al. 2018; Genus and Stirling 2018; Mazzucato 2018; Schot and Steinmueller 2018).

Very broadly, arguments against any notion of 'purity' in relation to the Republic of Science norms and governance structures calls for university researchers, and the SGCs which support them, to embed themselves as engaged actors working directly and closely with others in the interests of social and economic development. Research funders must enable this embeddedness (AAS 2018; Arocena et al. 2018). These perspectives coincide with critical assessments of the power relations embedded in high degrees of scientific autonomy. Science, technology and society (STS) scholars such as Andrew Stirling and Brian Wynne have analysed the power structures related to autonomy from the perspective of the privileged position that it gives scientists and a scientific elite (Stirling 2007, 2014; Wynne 2007, 2010).

Whilst the case against an ivory tower mentality is extremely strong, critics often ignore important political economy dimensions in debates about scientific autonomy. Whereas STS arguments pertain to the issue of autonomy and control in relation to scientists, there are other facets to the various framings and complex debates around scientific autonomy that are too easily ignored. The issue of autonomy for scientists is often treated as one in which, in the interests of efficacy and justice in science funding, influential academic knowledge-producing actors need to acknowledge the credibility and legitimacy of others.

However, whilst the Republic of Science portrays a world dominated by merit and reason, academic knowledge producers do not share power equally. Rather than one pure Republic of Science, which those striving for relevance have to reign in, the view from low- and middle-income countries (LMICs) national-level research environments is often that it is the lack of effective autonomy for researchers and SGCs aiming to fund academic research at national level which inhibits productive engagement.[7] International research collaborations and international funders, looking for high-profile research publications relating to the scientific frontier, skew prioritisation (Chataway et al. 2019). Rather than a straight trade-off between a cohesive collective of scientists on the one hand and policy-makers on the other hand, the issue of autonomy from this perspective relates to the degree of space that national-level actors have.

For example, in our recent study of SGCs, interviewees from SGCs and researchers themselves framed the issue of autonomy in different ways. In one framing, lack of political and economic space and resources was seen by a number of interviewees in different East African countries as a problem for national science funders (Chataway et al. 2019) who have fragile and compromised capacities to define agendas, which are truly in the public interest in SSA countries. Lack of various sorts of capital (social, political and economic) can inhibit effective operation and engagement between scientists and broader society at the national level. Low levels of political, economic and social capital and space for autonomy limits the extent to which scientists and science funders can engage effectively with policy-making communities and with international counterparts.

The problem of retaining capabilities to make local decisions about science, based on the relevance of expertise generated, is therefore partly to do with an ability to resist 'capture' by international conventions and establishments (Tilley 2011; Beigel 2013; Roy 2018). A recent article in *Nature* (Nordling 2018) discusses some of these issues in relation to the decolonisation of education, curriculum and research, using South Africa as the illustrative case. An evaluation of European Commission funding for research and development (R&D) for Poverty-Related and Neglected Diseases (PRND) revealed a widespread feeling amongst researchers that research conducted by international partnerships was often based on targets and priorities that limit the extent to which such research impacted on healthcare research partners in LMICs (Cochrane et al. 2017). A study by Pouris (2017) seems to confirm this finding.

The issue of lack of autonomy runs deep and includes different capabilities and capacities in the production and use of ST&I data and indicators which would allow SGCs to argue their corner more effectively (Manyuchi and Mugabe 2018), determine the direction of science and research, and play a leadership role in setting research agendas in SSA.

There are of course numerous and well-known examples of the damage that can result from extreme cases where scientific agendas correspond more to national political power than to rigour and excellence. Strong arguments are made that while there may be different ways of configuring SGCs in relation to strategic autonomy (Cruz-Castro and Sanz-Menéndez 2018), operational autonomy must be protected in more absolute terms.

Another dimension to the need for a degree of autonomy may rest on the ability of SGCs in SSA or regional and international research funders[8] to promote alternatives to dominant scientific and innovation trajectories. Current initiatives relating to the momentum behind calls for research funding to support transformative innovation experiments and mission-oriented approaches (Schot and Torrens 2017; Mazzucato 2018; Schot and Steinmueller 2018) argue that leadership needs both to be demand and user-led, but also have the ability to break with convention and avoid capture either by existing powers or regime actors or by existing convention (Russell 2015).

Thus, the challenge of constructing research agendas in ways which serve social, economic and environmental agendas raises a multitude of interesting and important questions about the relationship and dynamics between researchers and funders in relation to embeddedness and autonomy (Evans 1995). In addition, it highlights the importance, in some contexts, that academics and other stakeholders have attached to autonomy (Algañaraz Soria 2013; Beigel 2013).

The preceding paragraphs indicate that actually there is not a simple trade-off between autonomy and the power of scientists on the one hand and relevance and embeddedness on the other hand. To be effective societal actors, academic researchers and the SGCs which support them need to engage, based on having political, economic and social capital and a degree of autonomy in national contexts.

What do SGCs and researchers need in order to fulfil multiple mandates?

The issues briefly addressed above warrant further discussions and deeper thought. But we suggest that the issues have some immediate and practical upshots for SGCs. As outlined above, there are dual needs to embed research in society and to build and retain a political space and economic recourse to secure a degree of independence, authority and the ability to foster knowledge that is truly relevant (Chataway et al. 2017a). We have made the argument that making progress in navigating this terrain is best not viewed as a straight trade-off in power between academics and non-academics, but as a more complicated acknowledgement for engagement underpinned by a variety of 'capitals' in relation to SGCs and academics, which underpin effective interaction. Power struggles within the Republic of Science may be as important in this regard as power relations between academic and non-academic actors.

National-level SGCs need the space and resources to foster research that engages local communities in multiple ways which embed science, research and innovation in the realities of local contexts (AAS 2018), while, at the same time, retaining autonomy to ensure scientific rigour, excellence and relevance in research practice (Russell 2015) and policy directions (Daniels 2017). This need is clearly articulated by SGCs and researchers in the study carried out by Tijssen and Kraemer-Mbula (2018) and similar findings in Chataway et al. (2017a and 2019).

Thus, a primary role of SGCs will remain in organising peer and expert reviews of research. Establishing operational autonomy to oversee the peer and expert reviews of research is widely seen as important in ensuring quality and rigour. In this, the legacy of Republic of Science thinking remains. Nevertheless, demands for broader indicators of excellence, so that the value of researcher, in relation to wide-ranging goals of fostering development of the research environment and in relation to the need for science and research to address societal challenges, also needs to be respected at national level.

The take-away from this first part of the chapter is that across contexts and different organisational and institutional set-ups (Cruz-Castro and Sanz-Menéndez 2018), SGCs are involved in a dual and ongoing process to establish in varying degrees their own operational and strategic autonomy on the one hand, and on the other hand, to embed themselves in broader policy processes and societal processes and narratives. This duality, and the multiple mandates that Kruss and colleagues (Kruss et al. 2016a) have written about, is reflected in the way in which SGCs support and evaluate research.

To be effective, SGCs require vision, alliances (social capital), economic resources (economic capital) and political support (political capital). The African Union (AU), the AU Development Agency (AUDA) (formerly NEPAD, New Partnership for Africa's Development) and initiatives such as the SGCI are working in a range of ways to support SGCs as they navigate this difficult terrain and forge new ways of working. One clear implication is that national science and research funders, such as the government, need to find ways to articulate their needs in relation to international funding. This is a crucial area and one that warrants more attention and further policy analysis and research (AAS 2018).

In many respects this conundrum is not new. However, changes in the framing of science and research policies and accompanying funding mean that researchers and the SGCs that fund them are looking for new ways to construct that balance. Since the 1990s, innovation systems have heavily influenced science policy and done much to highlight the wide variety of institutions, organisations and intermediaries necessary to relate research to science. There are now growing demands that policy bodies and funders pay more attention to the direction of research so that it contributes in broader ways to social and environmental goals and economic well-being, as well as more conventional industrial connections (Stirling, 2007, 2014; Schot and Steinmueller, 2018).

One way to achieve this goal of ensuring that science and research address societal challenges could be through the inclusion of those traditionally considered to be 'non-scientists', for example, civil society groups and the private sector, in the formulation and implementation of relevant science and research projects. A broader group is also essential to achieving national innovation and development agendas (Daniels et al. 2017). Although the involvement of other groups in innovation, development and policy processes raises additional capacity, coordination, management and various other challenges for SGCs, this approach provides one avenue to addressing the complaints raised around the (mis)alignment of science and research to societal challenges in SSA. In line with our previous argument, however, this combination of academic and non-academic perspectives needs to be based on genuine engagement and attempts to co-construct agendas.

This has led to SGCs in many countries, including Colombia, Finland, Sweden, Norway, Japan and South Africa amongst others, making decisions to better align their funding to a range of local social and environmental policies, as well as industrial and growth goals. In some cases, Sweden for example, funding for innovation-related research is now explicitly linked to the Sustainable Development Goals (SDGs), while Colombia has recently produced a post-conflict ST&I strategy, a Green Paper that focuses on the SDGs and Transformative Change, underpinned by innovation policy (Chataway et al. 2017b; Schot et al. 2017). In the UK, the impact component of the Research Excellence Framework (REF) requires academics to develop case studies showing how their research contributes to non-academic goals. Although this approach is not linked to predefined social goals, it institutionalises a demand for all research departments (although not every academic) to relate their work to addressing societal challenges more broadly. There may be lessons for SGCs in SSA to draw from the UK's REF approach.

The need to broaden our frames of reference for engagement between researchers and society is echoed in many quarters, including from those working within innovation systems schools of thought that have previously focused on economic growth and links between industry and university (Fagerberg 2018). Lundvall (2007) highlights the fact that innovation systems approaches have been more useful for explaining the evolution of innovation systems than system building because of the largely unplanned and spontaneous nature of system evolution. Lundvall's argument stresses the reality of difficult living conditions in low-income countries which constrain people's ability and willingness to engage in work-based learning and participate in formal innovation processes. Against this backdrop, an obvious policy strategy is to target the wider context of the innovation system in such a way as to reduce these difficulties by, for example, enhancing stability, basic living conditions and access to basic services. This needs to be done in tandem with more conventional efforts to enhance scientific and technological capabilities, as well as institutional and organisational capabilities. SGCs in SSA can potentially play important roles in forging links across policy domains.

The following section considers some new approaches being implemented by SGCs in fostering new ways of connecting science and research to addressing societal goals, and connecting researchers to the broader society.

Navigating Republics and embedding excellence

One way in which research funders and researchers have sought to fund research relevant to local contexts is to fund 'applied science'. In the political economy study that the SPRU/ACTS team carried out, applied science emerges as a priority for all SGCs in case study countries (Chataway et al. 2017a). What was less clear is what was meant by applied science and how applied research was differentiated from basic science. This lack of clarity was compounded by the fact that public sector funding for applied work did not seem to be related to networks including private sector or civil society actors. As far as our evidence allowed us to judge, there seemed to be very few instances of applied funding. This begs the question, 'applied to what?'.

More broadly, questions about the usefulness and legitimacy of the distinction between applied and basic science have been raised by science policy analysts for some time (Calvert 2006; Narayanamurti and Odumosu 2016). Calvert (2006) for example suggests that the categories are used in fairly random ways as devices to generate support for particular initiatives. Narayanamurti and Odumosu (2016) on the other hand, writing in the context of the United States of America, argue that separating science into the two broad categories of 'basic' and 'applied' is a false distinction, and that this distinction limits science/research and hinders policy.

For SGCs, it could be useful to view the underlying need to support relevant research from a process and capabilities standpoint. Rather than providing support to a category of research labelled as applied, SGCs need to support a range of capabilities that will enhance capacities to generate and diffuse socially relevant science and research. Capabilities are also essential if SGCs are going to get better at conceptualising science, research and innovation in ways that ensure embeddedness or relevance, and shape key policy directions in Africa (AOSTI 2013; Daniels 2017).

To be effective, these capability-building efforts need to be related to research supported by stakeholder engagement exercises. This perspective highlights the importance of achieving relevance by means of different stakeholders being able to engage in a process around collective development of science and research agendas, broad-based consultations during research, and potentially carrying out research jointly; that is, involving multiple stakeholders in an inter- or transdisciplinary manner. This generates different sorts of 'capital' in Bourdieu's terms (Russell 2015) and capabilities relating to identified objectives (Chataway et al. 2017b; Schot and Steinmueller 2018).

Whilst many would argue that it is critical for SGCs themselves to retain control over the review process and with regard to final decisions about how and what to fund, we stress the need for participation and broader stakeholder engagement. A variety of studies point to the value of having engagement in formulating and carrying out research based on the following criteria (Russell 2015):

- Normative (from a power and justice point of view, to encourage participation offers a chance for non-academics to engage with an area that they are funding through taxes);
- Instrumental (it is more likely that research will have societal relevance if it is based on the engagement of different actors); and
- Epistemological (the ability to create knowledge communities which are able to develop new pathways, and approaches to relate science and society are enhanced by new communities and ebb and flow in social capital).

So, an approach that recognises the importance of a range of different capacities and capabilities in order to achieve goals is necessary. This approach highlights the importance of funding not only discrete research projects, but also funding networking and engagement activities designed to facilitate conversations between researchers, government ministries, civil society actors, a range of private sector bodies and civil society stakeholders. Responsive mode calls may not require these forms of engagement, but funding mechanisms that are designed to encourage research relevant to more immediate aims are likely to benefit from efforts to increase engagement. Engagement exercises can be in relation to particular challenges or broad issues and extend SGCs' remits beyond only academically valued research or boundaried public–private partnerships (PPPs) to a broader remit of supporting research and engagement activities (Palmberg and Schwaag Sherper 2017). This broader remit could improve the prospects of research that better contributes to addressing societal challenges and perhaps underpins broader approaches to thinking about excellence.

Engagement exercises and research based on stakeholder engagement can be used as part of inter- and transdisciplinary exercises in numerous ways. The following are a few examples:

- Exploration of ways to 'ground', contextualise and sense-make scientific research. For example, positive results from clinical trials to assess the effectiveness of antiretroviral drugs (ARVs) in preventing as well as treating HIV/Aids was received in radically different ways according to ability and desire to integrate new treatment options into existing treatment pathways and policies. An engagement exercise around the results helped clarify the implications of clinical trial results and define options for policy-makers and health systems decision-makers (Morgan Jones et al. 2014). This is just one example, but there are numerous others which might be proposed if SGCs design funding calls constructed to enable researchers to explore how best to make use of recent scientific developments. This type of approach is one way of aligning local research agendas with developments at the 'global frontier'. It does not of course overcome the issue of how local research spending can be skewed by international research funding patterns.
- Calls based on research partnerships and the co-creation of research are increasingly common. For example, (1) partnerships for vaccine development in relation to capacity building in health and innovation (Hanlin 2008); or (2) joint research chair

> initiatives, in which the IDRC has ample experience and has collaborated with various actors in developing countries. In these examples, the partnerships, collaborations or research chair initiatives help to build capacity, focus on research which is relevant to the countries involved and foster development. These initiatives, which sometimes take the format of PPPs, are often thought of as useful ways across many contexts to link research and development (Hanlin 2008; see also Oyelaran-Oyeyinka et al. (2018) for a summary). However, evaluations often underscore the need for national public sector partners, including research partners, to have adequate resources, capabilities and capacities (Marjanovic et al. 2015; Eurodad 2018) and to be able to deploy their various 'capitals' with operational autonomy. Although there is not any clear evidence that PPPs always lead to good outcomes, the examples above highlight where good capacity-building outcomes have been achieved within specific contexts, resulting in strengthening of health systems.

In a number of contexts the SDGs have inspired or are being used to structure new approaches to science funding and support in national contexts. For example, drawing heavily on the work of the Transformative Innovation Policy Consortium (TIPC) (Chataway et al. 2017b; Schot et al. 2017), Colombia is proposing to restructure its science funding around transformative innovation (El Libro Verde 2018). Whether or not these initiatives prove successful will need to be monitored and evaluated, but they represent powerful examples of experiments in funding research, which drive science in particular directions based on assessment of social, environmental and economic needs.

The desire of academics to work on these types of embedded research approaches may well depend on the way their work is evaluated (Kruss et al. 2016b) and the impact that engaging in interdisciplinary and transdisciplinary work has on academic careers. This takes us back to questions about indicators and metrics and research evaluation, and directly to how different versions of excellence are valued (Wilsdon et al. 2005, 2015).

Concluding thoughts: New approaches for embedded excellence

This paper has looked at different dimensions of the debate over scientific autonomy and discussed the need for funding and supporting research, which both reflects a respect for scientific excellence and embeddedness (or relevance, quality). In achieving this excellence versus quality objective, there is the need to extend the definition of excellence in ways that embed research in social, political, economic and policy contexts. This notion of embeddedness therefore constitutes the key argument and contribution that this chapter seeks to make.

In developing this notion of embeddedness, we have discussed some of the 'capitals', capabilities and capacities needed to support the process of embedding excellence at the national level. This includes new national and international understandings of the ways in which different sorts of research agendas can create support and synergies with each other. In addition, we point out that realising the desired level of embeddedness will require the aligning and realigning of national and international science and research agendas and funding, across different sectors and systems of critical development importance.

Furthermore, we have argued that the process of embedding excellence requires expanding the criteria for assessing quality and for science and research to have direct relevance to pressing national-level social, economic and environmental and policy issues. In order to achieve this objective, SGCs will have to do a number of things, which includes: (a) take greater ownership of their science and research agendas; (b) exercise higher levels of autonomy in their activities and decision-making; and (c) design and implement science and research projects, and funding schemes, in ways that encourage the involvement of non-academic actors. In doing this, SGCs also have to accumulate and deploy their various sources of strength and capital to make sure that research is seen to be trustworthy (i.e. maintaining scientific rigour and excellence), while remaining relevant to societal goals and needs.

We have outlined some of the thinking which underpins the notion of embedded excellence as an alternative to the notion of excellence based on traditional indicators and metrics, such as publications, or the distinction between applied and basic research. In the later part of the chapter, we developed ideas on how SGCs, and research councils in general, can more strategically orient themselves in the context of the above critiques and apply some of the practical suggestions in the chapter. Finally, we provided some practical suggestions in which the concept of embedded excellence might guide the work of SGCs in SSA, but also science councils elsewhere.

Notes

1 For more on SGCs, see Chataway et al. (2017a) *Case Studies of the Political Economy of Science Granting Councils in SSA*. Available at: https://sgciafrica.org/en-za/resources/Resources/PoliticalEconomy.pdf and Chataway et al. (2019)
2 For the missions see: https://sgciafrica.org
3 This chapter builds on a study undertaken for SGCI on the Political Economy of SGCs in SSA (see Chataway et al. 2017, 2019).
4 Other chapters in this volume look in detail at the perverse consequences of journal impact factors; therefore we do not focus on those arguments here.
5 Ideas about science policy and David Goodhart's book *The Road to Somewhere* can be found here: http://tipconsortium.net/science-and-innovation-policy-as-though-somewhere-mattered/
6 Daniel Sarewitz (2017) Saving science: Science isn't self-correcting, it's self-destructing. To save the enterprise, scientists must come out of the lab and into the real world. *The New Atlantis*. https://www.thenewatlantis.com/publications/saving-science
7 In Caroline Wagner's terms as laid out in *The New Invisible College*, the calls for link and sink strategies whereby national level investment accompanies strategies aimed at international integration.
8 Regional funders are e.g. Alliance for Accelerating Excellence in Science in Africa (AESA), while international funders include the likes of the UK's Wellcome Trust and DFID, the Gates Foundation and World Bank (for more on this, see e.g. Chataway et al. 2018).

References

AAS (African Academy of Sciences) (2018) Africa Beyond 2030. Leveraging Knowledge and Innovation to Secure Sustainable Development Goals. Nairobi: AAS

Algañaraz Soria VH (2013) Between scientific autonomy and academic dependency: Private research institutes under dictatorship in Argentina (1976–1983): The case of FLACSO. In: F. Beigal (ed.) *The Politics of Academic Autonomy in Latin America.* Surrey: Ashgate

AOSTI (2013) Science, technology and innovation policy-making in Africa: An assessment of capacity needs and priorities. *AOSTI Working Paper No. 2*

Arocena R, Göransson B and Sutz J (2018) *Developmental Universities in Inclusive Innovation Systems Alternatives for Knowledge Democratization in the Global South.* Palgrave Macmillan

AUC (2014) *Science, Technology and Innovation Strategy for Africa 2024.* Addis Ababa: African Union Commission

AUC (2015) *Agenda 2063: The Africa We Want. A Shared Strategic Framework for Inclusive Growth and Sustainable Development: First Ten-Year Implementation Plan 2014–2023.* Addis Ababa: African Union Commission

Beigel F (2013) *The Politics of Academic Autonomy in Latin America.* Surrey: Ashgate

Benner M (2011) In search of excellence? An international perspective on governance of university research. In: B Göransson and C Brundenius (eds) *Universities in Transition: The Changing Role and Challenges for Academic Institutions.* London: Springer. pp. 11–24

Bush V (1945) Science: The endless frontier. *Transactions of the Kansas Academy of Science (1903-)* 48(3): 231–264

Calvert, J (2006) What's special about basic research? *Science, Technology & Human Values* 31(2): 199–220

Chataway J, Ochieng C, Byrne R, Daniels C, Dobson C, Hanlin R et al. (2017a) *Case Studies of the Political Economy of Science Granting Councils in sub-Saharan Africa.* Report for the Science Granting Council Initiative. *https://sgciafrica.org/en-za/resources/Resources/PoliticalEconomy.pdf*

Chataway J, Daniels C, Kanger L, Ramirez M, Schot J and Steinmueller E (2017b) *Developing and Enacting Transformative Innovation Policy: A Comparative Study.* http://www.tipconsortium.net/wp-content/uploads/2018/04/Developing-and-enacting-Transformative-Innovation-Policy-A-Comparative-Study.pdf

Chataway J, Dobson C, Daniels C, Byrne R, Hanlin R and Tigabu A (2019) Science granting councils in sub-Saharan Africa: Trends and tensions. *Science and Public Policy* 46(4): 1–12. https://doi.org/10.1093/scipol/scz007

Cochrane G et al. (2017) *Evaluation of the Impact of the European Union's Research Funding for Poverty-Related and Neglected Diseases. Lessons from EU Research Funding (1998–2013).* Published in: EU Law and Publications (September 2017). doi: 10.2777/667857

Cruz-Castro L and Sanz-Menéndez L (2018) Autonomy and authority in public research organisations: Structure and funding factors. *Minerva* 56(2): 135–160. https://link.springer.com/article/10.1007/s11024-018-9349-1

Currey S (2017) Must science be useful? *The New Atlantis* Summar/Fall

Daniels, C (2017) Science, technology and innovation in Africa: Conceptualisations, relevance and policy directions. In: C Mavhunga (ed.) *What Do Science, Technology and Innovation Mean from Africa?* Chicago, USA: MIT Press

Daniels C, Ustyuzhantseva O and Yao W (2017) Innovation for inclusive development, public policy support and triple helix: Perspectives from BRICS. *African Journal of Science, Technology, Innovation and Development*: online. https://doi.org/10.1080/20421338.2017.1327923

De Saille S (2015) Innovating innovation policy: The emergence of 'Responsible Research and Innovation'. *Journal of Responsible Innovation* 2(2): 152–168

El Libro Verde (2018) *Green Book 2030: Science and Innovation Policy for Sustainable Development.* Colciencias, Colombia

Eurodad (2018) *History RePPPeated: How Public-Private-Partnerships are Failing.* https://eurodad.org/files/pdf/1546956-history-repppeated-how-public-private-partnerships-are-failing-.pdf

Evans P (1995) *Embedded Autonomy: States and Industrial Transformation.* Princeton: Princeton University

Fagerberg J (2018) Mobilizing innovation for sustainability transitions: A comment on transformative innovation policy. *Research Policy* 47(9): 1568–1576

Flink T and Kaldewey D (2018) The new production of legitimacy: STI policy discourses beyond the contract metaphor. *Research Policy* 47(1): 14–22

Genus A and Stirling A (2018) Collingridge and the dilemma of control: Towards responsible and accountable innovation. *Research Policy* 47(1): 61–69

Hanlin R (2008) Partnerships for vaccine development: Building capacity to strengthen developing country health and innovation. PhD Thesis, University of Edinburgh

Kruss G, Haupt G, Tele A and Ranchod R (2016a) *Balancing Multiple Mandates: The Changing Roles of Science Councils in South Africa.* HSRC Publishing

Kruss G, Haupt G and Visser M (2016b) 'Luring the academic soul': Promoting academic engagement in South African universities. *Higher Education Research and Development* 35(4): 755–771

Lundvall, B-A (2007) Innovation System Research and Policy: Where it came from and where it might go: online. http://www.globelicsacademy.org/2011_pdf/Lundvall_(post%20scriptum).pdf

Manyuchi AE and Mugabe JO (2018) The production and use of indicators in science, technology and innovation policy-making in Africa: Lessons from Malawi and South Africa. *Journal of Science and Technology Policy Management* 9(1): 21–41. https://doi.org/10.1108/JSTPM-06-2017-0026

Marjanovic S, Cochrane G, Manville C, Harte E, Chataway J and Jones MM (2015) *Leadership as a Health Research Policy Intervention: An Evaluation of the NIHR Leadership Programme (Phase 2).* Santa Monica, CA: RAND Corporation. https://www.rand.org/pubs/research_reports/RR934.html

Mazzucato M (2018) *Mission-oriented Research & Innovation in the European Union. A Problem-solving Approach to Fuel Innovation-led Growth.* European Commission

Morgan Jones M, Castle-Clarke S, Brooker D, Nason E, Huzair F and Chataway J (2014) *The Structural Genomics Consortium: A Knowledge Platform for Drug Discovery.* Santa Monica, CA: RAND Corporation. https://www.rand.org/pubs/research_reports/RR512.html

Narayanamurti V and Odumosu T (2016) *Cycles of Invention and Discovery: Rethinking the Endless Frontier.* Cambridge: Harvard University Press

Nordling L (2018) South African science faces its future. *Nature*, 7 February

Oyelaran-Oyeyinka B, Vallejo B, Abejirin B, Vasudev S, Ozor N and Bolo M (2018) Towards Effective Public–Private Partnerships in Research and Innovation: A Perspective for African Science Granting Councils. *African Technology Policy Studies Network (ATPS) Technopolicy Brief No. 49*

Palmberg C and Schwaag Scheper S (2017) Towards next generation PPP models – insights from an agency perspective. Conference paper. https://www.researchgate.net/publication/315713974_Towards_next_generation_PPP_models_-_insights_from_an_agency_perspective

Polanyi M (1962) The Republic of Science. *Minerva* 1(1): 54–73

Pouris A (2017) The influence of collaboration in research priorities: The SADC case. *South African Journal of Science*. 113(11/12): online. http://dx.doi.org/10.17159/sajs.2017/20170150

Rip A (1994) The Republic of Science in the 1990s. *Higher Education* 28(1): 3–23

Roy RD (2018) Decolonise science – time to end another imperial era. *The Conversation*, 27 June

Russell LD (2015) Democratizing the scientific space: The constellation of new epistemic strategies around the emerging metaphor of socially embedded autonomy. *Technology in Society* 40: 82–92

Schot J and Steinmueller WE (2018) Three frames for innovation policy: R&D, systems of innovation and transformative change. *Research Policy* 47: 1554–1567

Schot J and Torrens J (2017) The Roles of Experimentation in Transformative Innovation Policy. *TIPC Research Brief 2017-02*

Schot J, Daniels C, Torrens J and Bloomfield G (2017) Developing a Shared Understanding of Transformative Innovation Policy. *TIPC Research Brief 2017-01*

Stirling A (2007) 'Opening up' and 'closing down': Power, participation, and pluralism in the social appraisal of technology. *Science, Technology, & Human Values* 33(2): 262–294

Stirling A (2014) Transforming power: Social science and the politics of energy choices. *Energy Research & Social Science* 1: 83–95

Tijssen R and Kraemer-Mbula E (2018) Research excellence in Africa: Policies, perceptions, and performance. *Science and Public Policy* 45(3): 392–403. https://doi.org/10.1093/scipol/scx074

Tilley H (2011) *Africa as a Living Laboratory, Empire, Development, and the Problem of Scientific Knowledge 1870–1950*. Chicago: University of Chicago Press

Wilsdon J, Allen L, Belfiore E, Campbell P, Curry S, Hill S et al. (2015) *The Metric Tide: Report of the Independent Review of the Role of Metrics in Research Assessment and Management*. DOI: 10.13140/RG.2.1.4929.1363

Wilsdon J, Wynne B and Stilgoe J (2005) *The Public Value of Science: Or How to Ensure That Science Really Matters*. Demos

Wynne B (2007) Public participation in science and technology: Performing and obscuring a political–conceptual category mistake. *East Asian Science, Technology and Society: An International Journal* 1: 99–110

Wynne B (2010) *Rationality and Ritual: Participation and Exclusion in Nuclear Decision-Making*. 2nd edn. London: Earthscan

CHAPTER 4

Re-valuing research excellence: From excellentism to responsible assessment

Robert Tijssen

Excellence and excellentism

Research excellence (RE) has become a very powerful concept in 21st century science policies. The *Oxford English Dictionary* defines 'excellence' as 'to be superior or pre-eminent, to surpass others'; it is a normative concept that acquires its meaning only in a proper comparative context. It is often presented as 'supreme quality' – a distinctive mark (the verb 'to excel' originated from the Latin verb *cellere* – i.e. to rise high). The fusion of 'excellence' and 'research' suggests an almost indisputable measure of quality, of being the best within a group of comparators. Within the area of science and scientific research, the notion has certainly caught on as referring to a desirably high level of performance. Nobel prizes are often considered, especially by the general public, to be an ultimate accolade of international excellence. High performance, excellent individuals or organisations are regarded as PR and marketing assets, which may not only attract wider attention in the press, but also boost research funding success rates.

It seems that every major city, region or country worldwide now aspires to have at least one centre of research excellence in its national science system, preferably housed prominently at the local university.

Any web search will show a proliferation of research organisations, university websites and science funding agencies that have tagged someone or something as excellent. But what is the quality of the evidence? It usually refers to some well-deserved prestigious award or noteworthy achievement, but more and more without convincing evidence to back up such a claim to fame (Sørensen et al. 2015). In the current hype and buzz, RE seems at risk of becoming a strategic construct that is ever more loosely connected to its originally intended meaning. This process of 'excellentism'[1] creates an environment in which excellence seems to be an increasingly easy target for misinterpretation and misuse. Some outspoken critics go so far as to describe the ongoing rhetoric as nothing less than fetish where RE has become a catch phrase in which performance has taken on almost mystical qualities (Moore et al. 2017).

Responsible assessment of research excellence

Similar to 'research quality', RE remains a fuzzy and unstable construct. And it is not difficult to see why: RE suffers from divergent theoretical perspectives, a plethora of analytical frameworks and a wide range of performance indicators (both quantitative and qualitative). Narrowly defined criteria of what quality RE may, or may not, entail are susceptible to criticism from those being assessed and may create fierce disputes between all parties involved. Some may say that, like any other subjective assessment, such assessment processes are bound to be messy and pragmatic, driven by incomplete information and shifting considerations.

Running an assessment system means facing many methodological challenges, analytical practicalities and implementation issues with regard to the required information to pass judgement. In addition to designing transparent protocols, checking data validity, ensuring sufficient comparability and many other concerns, one must also choose the most appropriate information items – opinions may differ widely as to how appropriate some of those selected items actually are.

Allowing access to understandable information is essential. Quality assessment inevitably involves an external review of relevant outputs. *Ex ante* assessments of research proposals, often describing anticipated research achievements, differ from information-gathering methodologies in *ex post* evaluations of research performance. Where proposal assessments tend to be based on the subjective opinions of individual experts or panels, thereby introducing the risk of questionable or unreliable information, evaluations are more likely to incorporate objectified data extracted from tangible outputs such as scientific publications. Research articles published in high-impact, peer-reviewed international scholarly journals, or books issued by international publishers, are usually recognised by the scientific community as significant *ex post* achievements.[2] But such outputs are no longer seen as the ultimate proof of quality; the focus has shifted to the appreciation by users of those impacts.

An increasingly large number of indicators-based analysts now prefer to operationalise and quantify RE in terms of producing high levels of citation impact within the international scientific community (Tijssen et al. 2002). Such a narrow definition of RE, reflecting knowledge creation outcomes of radical novelty, presents an extremely homogenised case of global RE. Some experts and scholars prefer to see research impacts, rather than outputs, as the defining part of research quality and apply impact-based standards to capture RE (OECD 1997; Boaz and Ashby 2003; Tijssen 2003). Other analysts note that quality and research impact are actually two different elements of research excellence (Grant et al. 2010).

To avoid the risk of becoming a truly contentious concept, and perhaps even a meaningless term, more transparency is needed. To achieve this, we should move away from a focus on research output or impact-related 'achievement-based' descriptions. RE should be more broadly framed, and transcend beyond the production of ground-breaking scientific discoveries and impacts of the global scientific community. RE is now usually viewed as being highly multi-dimensional and can manifest itself in different ways and at various stages of research processes: across a wide range of 'input' dimensions (originality of research proposals, human capital development, research infrastructures, etc.); but also via 'throughputs and processes' (methodological rigour, ethics compliance, reproducibility, etc.);

'outputs' (ground breaking, internationally leading, etc.); and impacts (scholarly, cultural, socio-economic). The focus on outputs is gradually being replaced by that of outcomes, in terms of their relevance and impacts, as a decisive indicator of high-quality RE.

Current science policies, mostly in Europe, have started to embrace this broader perspective. Acknowledging a multidimensional view, the overarching notion of *Responsible Research and Innovation* (RRI)[3] is becoming one of the major driving forces in ongoing debates on the future of science. In the broader framework of RRI, research performance incorporates a range of good scientific practices, such as 'open access' publications and 'open science' data sharing, ethical considerations and societal responsibility. RRI-driven assessments of research performance should develop more appreciation for interdisciplinary research and aim to open up new dimensions of scientific quality – not only with regard to application-oriented (or applied) science and social innovation by practitioners, societal engagement with policy-makers and the public, but also for the representation of minorities in the scientific community.

Research excellence in the Global South

Aspirations and initiatives to achieve 'research excellence', without any clear definition of the core concept and how it should be operationalised in performance assessments, are likely to produce misguided policies and sub-optimal investments. In an era where many public sector science budgets are no longer increasing, and tough choices about funding priorities are unavoidable, we need more clarity on the merits of RE-guided policy initiatives. This predicament applies full force to low- and medium-income countries (LMICs) in the Global South, especially in those countries that aim to catch up or benchmark themselves with the world's scientific leaders. Where science budgets are low and aspirations are high, LMICs tend to emulate science policy models and associated research assessment systems from the Global North. In doing so, not only do they run the risk of ignoring local societal needs, but also of downplaying the existence of relevant indigenous research strengths.

Science funding and RE ambitions in the Global South require a customised approach (Tijssen and Kraemer-Mbula 2018). To gain more clarity on if and how investments in science are delivering sufficient value for (inter)national funders of science, a more focused discourse is needed on establishing productive meanings of RE and associated concepts. Tijssen and Kraemer-Mbula (2017), in their policy brief entitled 'Perspectives on research excellence in the Global South: Assessment, monitoring and evaluation in developing-country contexts', present a critical view of mainstream methodologies to assess and evaluate RE in African science systems. The policy brief proposes practical suggestions for more appropriate analytical models and diagnostic kits, geared towards the needs of science funders and review panels that inevitably operate in difficult, resource-constrained policy environments. One of the brief's main general conclusions states that

> evidence-based decisions on science funding require robust science policy tools and analytical frameworks. Future contributions could consider different avenues and perspectives that can help science granting councils around the world, but especially in the LMICs of the Southern hemisphere, address a perceived need to fund research excellence without sacrificing broader objectives related to research impact, inclusivity, social responsibility, transparency and accountability.

To develop and implement such instruments, one first needs to recognise and acknowledge that any attempt to clarify or harmonise RE's multidimensionality runs into a set of 'wicked'[4] conceptual and methodological problems. The remainder of this chapter picks up on where the above-mentioned policy brief ended, namely with the following two research questions to guide practical steps to re-value RE:

- Is RE an appropriate objective for research-funding decisions in the Global South?
- Which RE-oriented analytical models, tools and assessment frameworks should be applied with the specific intent of strengthening local research?

Focusing specifically on LMICs (African countries in particular), these questions require further critical thinking and empirical analysis. Discussing the topic of RE within a problem-driven, interdisciplinary context, one faces idiosyncratic logics and conflicting views that force evaluators, analysts and stakeholders to justify what we are doing and why. Core assumptions and expectations about the nature of RE and its impacts will inevitably differ. So, we need to ask ourselves the underlying question: do we need to develop a shared understanding of RE, and if so why? The next subsections will present information and arguments to answer this core question affirmatively.

Conceptual issues and methodological problems

Any effective discourse and decision-making on how to perceive RE should be driven by shared terminology and common definitions. A generally accepted 'dominant' heuristic is needed to help identify RE in its many shapes and forms; a convincing rhetoric is required to influence researcher communities and their major stakeholders. Only then can one hope to arrive at a set of methodological principles that can underpin common practices with regard to the assessment of research proposals, activities and outcomes.

The late Robert Merton – one of the founding fathers of the sociology of science[5] – presents a plea for more clarity on the topic, apparently driven by the reluctance he observed in this environment to pin down the key characteristics of research achievements and the associated notion of excellence:

> Many of us are persuaded that we know what we mean by excellence and would prefer not to be asked to explain. We act as though we believe that close inspection of the idea of excellence will cause it to dissolve into nothing. (Merton 1973: 422)

Merton poses three pivotal questions to aid us in a closer examination of RE:

- What unit of achievement is to receive recognition?

- Who shall judge the achievement?
- What qualities of achievement are to be judged?

As for the first question, some examiners of research proposals or evaluators of achievements will argue that RE is primarily about the individual researcher as a unit of assessment. Striving for RE, or attaining it, is then about personalised processes of creativity, methodological rigour and achievement. Those are the gifted individuals who are able to create new knowledge and innovate. Such 'excellent' researchers are the ambassadors of rich and diverse science ecosystems with 'research cultures' that are diverse, innovative and quality driven. Adopting this micro-level, person-oriented viewpoint, organisations or networks can never be regarded as excellent. Fine-tuned and tailored incentive systems become essential conditions for RE, as well as dedicated human resource management practices and researcher-centered performance assessment systems. Another strand of evaluators might stress the importance of organisational factors, external determinants and accumulated earlier achievements by others. Although RE is still seen as a person-embodied level of performance, it is now primarily facilitated, shaped and driven by environmental, organisational and historical circumstances and developments. The organisation is the main unit of achievement.

Irrespective of how RE is perceived or at what level it is assessed, of particular interest remains the extent to which outstanding scientific achievements are recognised and judged in accordance with common understandings of quality, relevance and impact. Clarity on these issues opens up the possibility to develop and apply assessment models and practices that target those research characteristics that are valued most within the context of science in LMICs.

Academic literature review

Focusing on Merton's third question, this section presents a summary review of academic studies to shed some light on how to create a clearer general understanding of RE within the context of research performance assessment frameworks. A comprehensive literature study, stretching back 50 years to Robert Merton's seminal work, does not exist. However, the scholarly literature of recent years shows a flurry of academic case studies on RE issues, usually within the context of evaluating university research performance, excellence-promoting policies within public science systems or the surge of centres of excellence. This contemporary review draws from the following 11 academic studies, all published in the international scientific literature: Laudel and Gläser 2014; Sørensen et al. 2015; D'Este et al. 2016; Ofir et al. 2016; Carli et al. 2018; Confraria et al. 2018; Ferretti et al. 2018; Fudickar and Hottenrott 2018; Moher 2018; Schmidt & Graversen 2018; and Tijssen & Kraemer-Mbula 2018. These studies address many of today's issues – often framed in science assessment and research evaluation settings – and provide several valuable new insights on topics of conceptualisation and operationalisation. The study by Tijssen & Kraemer-Mbula is specifically targeted at the situation in Africa.

With regard to Merton's question as to 'what qualities of a seeming achievement are to be judged?', Laudel and Gläser (2014) stress the value of peer review to assess RE:

> The properties used to characterise exceptional research ('major discovery', 'creativity', 'breakthrough') are extremely vague, and are not operationalised for empirical identification either. This is why the major studies addressing conditions for that research let the scientific communities decide which of its research was exceptional and then studied conditions for this research. (Laudel and Gläser 2014: 1205)

However, some studies also highlight features of RE that are measurable, such as:

> The results of a number of previous studies which focused on the relation between expert panel assessments and quantitative assessments, such as bibliometric outcomes of research units, reveal that assessments of expert panels are positively related to publication and citation indicators. (Schmidt & Graversen 2018: 359)

> An alternative to using counts and rankings of awards and prizes, which we pursue in this study, is to identify awarded (or funded) scientists as a comparison group and then to use their publication records and project description content for science evaluations. This approach provides us with an external 'reference point' or knowledge frontier, to which we can compare other scientists. (Fudickar and Hottenrott 2018: 6)

Other studies emphasise the importance of teamwork and cooperation to achieve excellence:

> Excellent knowledge embedded in researchers and research teams can also be measured through research grants. The most prominent (high value and prestige) research grants, such as that of the European Research Council (ERC) or the National Science Foundation (NSF) of the United States are awarded based on demonstrated outstanding past performance of research teams on the one hand, and on expected outstanding performance on the other hand. Receiving such a grant can therefore be at the same time a proxy for recent excellence and 'excellence in the making'. (Sørensen et al. 2015: 229)

> [I]n this study, we assumed that any co-author of a highly cited paper made a significant contribution to that paper. However, it has been suggested that researchers in lower-income contexts are rarely leading authors in international publications and that their role is often still primarily limited to collecting data and linking up with domestic policy debates. (Confraria et al. 2018: 230)

> According to the views of surveyed SGC research co-ordinators, current legal frameworks still constitute a developmental challenge since they do not explicitly foster the pursuit of research quality involving research collaboration networks (national and international, among

> researchers and with users/stakeholders). As a result, a 'silo mentality' often prevails in African research performance, which is seen as a major deterrent to achieve RE. (Tijssen & Kraemer-Mbula 2018: 402)

While, finally, several authors perceive the research environment and user communities as major determinants:

> Overall, our results showed that individual features influenced research excellence, but that context also played a fundamental role. [...] Contextual variables reinforced individual performance: if an academic works in an environment to which other excellent scholars are affiliated, a general research enhancement occurs, which is also sustained by the heterogeneity of the research setting. Conversely, if the work context is populated by academics with poor publication experiences, that would result in lower research standards. Finally, the quality of the research context moderated individual ability, in that an academic without a robust past research experience strongly benefited from a well-developed work setting that offered outstanding publication exposure. (Carli et al. 2018: 13)

> [T]he importance and value to key intended users of the knowledge and understanding generated by the research, in terms of the perceived relevance of research processes and products to the needs and priorities of potential users, and the contribution of the research to theory and/or practice. (Ofir et al. 2016: 10–11)

> What counts as excellence is entertained by the imagination of some about what 'excellent research' is; but what political, social, and ethical commitments are built into the adopted notion and the choice of what needs to be quantified? (Ferretti et al. 2018: 733)

The above ideas, suggestions and observations not only acknowledge a multitude of views and analytical approaches, but also reiterate that RE – a normative concept at its very core – is very much an integral part of complex social systems that require a much better understanding in order to design appropriate models and tailored assessment systems of scientific performance and RE.

Towards a better understanding

To achieve more clarity, preferably with solid empirical underpinnings, one should start by accepting that a consensual, working definition of RE is not likely to emerge very soon; as one solicits a wider range of inputs and views in a consultation process, a multitude of fundamental differences in ideas and perceptions will come to the fore. However, some degree of consensus on practical issues should be attainable. The collective intelligence from experts, as exemplified in the quotes above, offers valuable insights and concrete suggestions on how to move forward further operationalisation, categorisation and measurement of RE dimensions.

Young (2015) introduces a helpful distinction between 'zero-sum excellence' and 'threshold excellence'. Where the former, narrow definition rests on the assumption that excellence is a limited resource distributed among researchers by competitive means, the latter broader definition is based on the assumption that excellence is unlimited and is defined by inherent qualit(y)(ies). The zero-sum case follows a winners-take-all logic that most of the funding instruments still apply: evaluation of proposals leads to a ranked list, for which a selection cut-off point is chosen. Only those who meet this threshold are funded and rewarded; the others lose. The rise of global RE, coupled with decreasing odds of success, creates stratification and selection processes where funding decisions favour the leading, established researchers and their vested interests. In such regimes, the rewards for attaining RE tend to be concentrated in top performers, despite the fact that differences between this first-tier 'elite' and lower tiers can be small and/or difficult to judge. In contrast, threshold excellence could have a success rate of 100%, provided the standards or criteria that the judges define as excellent are met, or 0% if all submitted cases are considered to be of insufficient merit or quality. Even the incumbent elite may fail to comply with the set criteria.

Where many RE assessment systems and practices still tend to favour distributions according to zero-sum excellence, the science granting systems of the Global South are better served by threshold excellence approaches. Applying a threshold criterion introduces a stable performance target, which is compatible with distributive justice arguments. Once the primary selection criterion has been met, it opens the door to legitimately include additional considerations, or targeted selection criteria (such as the UN Sustainable Development Goals), to guide final decision-making on funding.

The International Development Research Centre (IDRC) conducted a study on how to evaluate research excellence, particularly of applied interdisciplinary research for development, to make the case for research that goes beyond generating new knowledge for the local or global scientific community (Ofir and Schwandt 2012). This study led to the Research Quality Plus (RQ+) model, developed by the IDRC, as a more holistic, practice-oriented approach to research evaluation (Ofir et al. 2016; Lebel and McLean 2018).

Indeed, from a stakeholder-based view, RE should be framed more explicitly in terms of research topics and capacities that address societal needs and collective interests. RE then becomes intertwined with relevance for, and impacts on, non-scholarly audiences and other user communities. The authors argue that RE is desirable in any type of research, but the stakes are higher when the outcomes are meant to influence decisions that affect people's lives, the environment, governance or other areas of development. Defining such 'local standards of excellence' in the resource-poor research environments of many LMICs, such calls for 'RE for development outcomes' should take into account local and domestic logistics and operational problems (funding, instrumental facilities, data capturing, software development, etc.). Implementing a stronger emphasis on local research issues and scholarship is essential to creating credibility of research outputs and impacts.

Towards operationalisation: Guiding principles and practical recommendations

Appropriate measurement of RE dimensions may improve the quality and effectiveness of research assessments and decision-making. But what is appropriate, and who determines that? How to pick the right kind of qualitative indicators or quantitative measures? The metrics marketplace is a confusing arena, with its variety of models and analytical tools. Applying indicators based on citation impact counts as a measure of impact in the global scientific community seems an obvious place to start, but there are many options available. Too often we are swayed by the availability of comparative quantitative data, such as the free online information in Google Scholar, rather than conducting a careful fit-for-purpose assessment of its analytical value. We tend to value what we can easily measure, rather than collecting empirical information on what is actually needed to characterise RE.

General guidelines to select and apply the most appropriate assessment toolkit have become necessary, especially with regard to use-oriented, applied research (McLean & Sen 2018). Fears that bibliometric indicators (i.e. performance measures based on publication output and/or citation impact) are being misused or misinterpreted, and are even damaging the system of research that they are designed to assess and improve, led to the publication of the *Leiden Manifesto for Research Metrics* (Hicks et al. 2015). The ten principles of this manifesto are: (1) quantitative evaluation should support qualitative, expert assessment; (2) measure performance against the research missions of the institution, group or researcher; (3) protect excellence in locally relevant research; (4) keep data collection and analytical processes open, transparent and simple; (5) allow those evaluated to verify data and analysis; (6) account for variation by field in publication and citation practices; (7) base assessment of individual researchers on a qualitative judgement of their portfolio; (8) avoid misplaced concreteness and false precision; (9) recognise the systemic effects of assessment and indicators; (10) scrutinise indicators regularly and update them.

All ten principles apply in full force to empirical studies and assessments of RE. Principal #3 is particularly important in Global South contexts, where high-quality scientific research on local issues or problems tends to be less visible and/or undervalued if it is not published in English or disseminated widely in international academic journals. Science in the Global South is often heavily involved in international research cooperation (Tijssen 2015), and empirical studies clearly show that high-impact research is dominated by trans-continental partnerships (Tijssen and Winnink 2018). Appropriate assessments therefore need to incorporate the contributions and impact of international research cooperation and networks.

Regarding principle #10, narrowly focused indicators are blunt instruments that may induce unfair comparisons of performances. Single-metric indicators, such as citation impact scores, can easily become 'metrics of mass destruction' when used in an uncritical, mechanistic fashion and driven by unrealistic performance targets.

The unit of assessment is important. High-level data that aggregate an entire continent, country, university or research institute, are seldom informative. Nor is micro-data information, on individual researchers, appropriate in most cases. The right level of granularity needs to align with how research is actually conducted: in small teams, organisational networks and in dedicated projects or coherent programs within LMICs (McLean and Sen 2018). High-quality 'responsible' assessments should target this intermediate level of detail and information content.

The above-mentioned 'Perspectives on research excellence in the Global South' policy brief (Tijssen and Kraemer-Mbula 2017; see subsection 1.3) presents ten methodological recommendations (see Box 1) that may help operationalise some of the guidelines into practical considerations and steps towards developing such a responsible RE-oriented assessment system.

Box 1. Recommendations for responsible assessment of research excellence

1. Science funders should be more explicit in their descriptions or definitions of 'research quality' and 'research excellence';
2. Determining 'excellence' is contingent on appropriate performance standards and benchmarks;
3. The appropriateness of a performance indicator depends on its degree of 'usability' and 'user acceptability' in terms of information value, operational value, analytical value, assessment value and stakeholder value;
4. Proper understanding and operationalising requires multiple perspectives (both local and global); it is important to make a clear distinction between common global benchmarks and 'local' customised ones;
5. Experiences within LMICs in adapting concepts of RE and 'research quality' to their local contexts constitute valuable sources of information to establish good practices in assessment and evaluation practices worldwide;
6. Expert opinions from peers should be a prime source of information for value judgements on research quality and excellence;
7. Personal views, usually embedded in implicit scientific norms regarding quality standards or driven by selected showcases of successful research, should be complemented by external empirical information to create 'informed peer-review' assessment and evaluation;
8. The multidimensional nature of research excellence requires an 'indicator scoreboard' approach, where performance indicators may span the entire spectrum from research resources to socio-economic impacts;
9. The choice of performance indicators and/or excellence benchmarks will always be context-dependent and goal-dependent; there is a clear need to incorporate local contextual factors in customised indicators;
10. Frameworks designed to assess research excellence ought to be flexible enough to incorporate changes in the local context and priorities, as well as in the dynamics of the global science system.

Source: Tijssen and Kraemer-Mbula (2017)

Broader perspectives for new approaches

RE assessment and evaluation approaches seem to be moving into a danger zone of ambiguity and unfulfilled potential. The current narrow focus, mainly on quantitative indicators of research outputs and scholarly impacts, needs a major rethink. A responsible way forward requires a broader scope of achievements and an upgraded analytical framework where research outcomes, societal impacts and user appreciation are key determinants. This upgrade implies mixed-method, multi-stakeholder assessments, based on tailored sets of indicators and a stronger focus on impact processes, while aiming to be as open (i.e. transparent, objective and fair) as possible for those being assessed or evaluated.

The recommendations in the box may help set the stage for (further) developing a set of guiding principles to implement an effective action plan. Consultation and mutual learning processes are an essential part of that development trajectory. We will have to accept that expert opinions and stakeholder consultations will not produce the same result, neither in terms of shared ideas or preferences on what RE entails nor in terms of key performance indicators (either qualitative or quantitative).

Such indicators are crucial for information gathering; collecting opinions is insufficient for high-quality comparative analysis and benchmarking. A thorough process of designing, testing and consolidating those indicators is equally important – to avoid narrowly defined 'one-dimensional' quantitative indicators of RE (such as the H-index) that may easily discard many other science-related achievements, or downgrade them to secondary criteria of research quality. Some indicators from the North, such as those applied in world university rankings, may exacerbate the gravitational pull towards homogenisation of research performance assessment, with its set of established metrics, and implicit 'knowledge hierarchies' that downplay the relevance of local contexts in which Global South universities and research institutes operate (Ndofirepi, 2017).

Applying RE-specific criteria and performance indicators developed for systems in the Global North is not advisable. Upgrading the research performance of scientists and scholars in the Global South cannot be accomplished within 'winners-take-most' funding systems that operate on 'the best versus the rest' selection mechanisms. Here we face a dilemma: adopting systems based exclusively on local standards is also not the best solution to create or promote RE in the Global

South. The outcomes of such one-size-fits-all policies and incentives will not always align with economic realities of science funders or their institutional expectations with regard to international excellence. Science funders should therefore take into account heterogeneity in the system and target different groups and contexts with appropriate interventions and customised assessment approaches. Implementing a strategic mix of international-level, zero-sum RE and local-level, threshold RE would avoid further increasing the level of heterogeneity within academic research systems. Developing such new assessment models and approaches will be well worth the effort. As Moher notes:

> How we evaluate scientists reflects what we value most and don't in the scientific enterprise and powerfully influences scientists' behaviour. Widening the scope of activities worthy of academic recognition and reward will likely be a slow and iterative process. The principles here could serve as a road map for change. While the collective efforts of funders, journals, and regulators will be critical, individual institutions will ultimately have to be the crucibles of innovation, serving as models for others. Institutions that monitor what they do and the changes that result would be powerful influencers of the shape of our collective scientific future. (Moher 2018: 16)

Returning to our core research question ('Do we need to develop a shared understanding of RE, and if so why?'), the main reason for an affirmative answer is that consensus on RE seems more urgent than ever in view of the emphasis of modern-day science as a major contributor to wealth and welfare in the local society. Tijssen and Kraemer-Mbula state that

> excellence is not only seen as a major marker of performance, but also as a driving force for forward-looking policies with high levels of political and organisational ambition. (Tijssen and Kraemer-Mbula 2018: 393)

It is therefore important and necessary that LMICs create their own sustainable research-intensive niches of excellence. What degree of concentration in research funding is optimal to create or sustain such niches? And how many dedicated resources are needed to upscale the 'stairway to excellence' to entire universities or research institutes? Both questions are difficult to answer and invite experimentation with research quality assessment systems to determine viable policy options. What is, after all, the point of implementing RE assessment systems if they don't contribute to responsible decision-making processes, evidence-informed research funding and performance assessments, and, ultimately, to a more beneficial and meaningful science?

Designing and implementing appropriate assessment systems not only requires enlightened mind-sets and a re-valuing of the conceptual framework, but also a rethink of criteria, protocols and procedures. Responding as much as possible to urgent local needs should be the prime consideration and key objective. To reconfigure and optimise such selection environments, funders and stakeholders should be willing and able to engage in experimentation and organisational learning processes. Science in the Global South deserves to have its own toolbox of good practices and performance indicators to support effective, forward-looking assessments of research excellence in all its richness and complexity. Africa needs an excellence culture where high-quality assessments support competent and credible scientific research.

Acknowledgement

Ms Lynn Lorentzen (CREST, Stellenbosch University) was helpful in searching the global scholarly literature for documents on 'research excellence'. This work is based on the research supported by the South African Centres of Excellence programme, funded by the National Research Foundation of South Africa (NRF Grant No. 91488). Any opinion, finding and conclusion or recommendation expressed in this material is that of the author and the NRF does not accept any liability in this regard.

Notes

1 This novel concept of 'excellentism' is meant to be a derogatory term to capture applications and mentionings of the term 'excellence' (or 'excellent') – usually in common language or the popular press – that lack any proper underlying definition or description, or grossly misrepresent, the essence of 'excellence' as a distinctive general concept.

2 Generating international impact may also involve high-profile presentations at academic meetings and forums, but also appearances before public audiences and a noticeable presence in social media.

3 The Rome Declaration on Responsible Research and Innovation in Europe states that 'the benefits of Responsible Research and Innovation go beyond alignment with society: it ensures that research and innovation deliver on the promise of smart, inclusive and sustainable solutions to our societal challenges; it engages new perspectives, new innovators and new talent from across our diverse European society, allowing to identify solutions which would otherwise go unnoticed; it builds trust between citizens, and public and private institutions in supporting research and innovation; and it reassures society about embracing innovative products and services; it assesses the risks and the way these risks should be managed (European Commission 2014).

4 'Wicked problems' (Churchman 1967) can never be satisfactorily solved in view of the underlying complexity of adaptive social systems that drive scientific funding and research.

5 Merton's work on scientific norms and his ideas one the nature of scientific production were also influential in the early development scientometrics and citation indices (see the chapter by Chavarro in this book).

References

Boaz A & Ashby D (2003) Fit for purpose? Assessing research quality for evidence based policy and practice. ESRC UK Centre for Evidence Based Policy and Practice: *Working Paper 11*. www.kcl.ac.uk/content/1/c6/03/46/04/wp11.pdf

Carli G, Tagliaventi M & Cutolo D (2018) One size does not fit all: The influence of individual and contextual factors on research excellence in academia. *Studies in Higher Education*. doi.org/10.1080/03075079.2018.1466873

Churchman C (1967) Wicked problems. *Management Science*. 14(4): B141–B142

Confraria H, Blanckenberg J & Swart C (2018) The characteristics of highly cited researchers in Africa. *Research Evaluation*. doi: 10.1093/reseval/rvy017

D'Este P, Ramos-Vielba I & Woolley R (2016) *Scientific and Social Impacts: Two Sides of the Same Excellence Coin for Individual Researcher Assessment?* Workshop 'Excellence policies in science' Leiden, 2-3 June 2016

European Commission (2014) *Rome Declaration on Responsible Research and Innovation in Europe*. Italian Presidency of the Council of the European Union. https://ec.europa.eu/digital-single-market/en/news/rome-declaration-responsible-research-and-innovation-europe

Ferretti F, Pereira A, Vértesy D & Hardeman S (2018) Research excellence indicators: Time to reimagine the 'making of'? *Science and Public Policy*: online. doi.org/10.1093/scipol/scy007

Fudickar R and Hottenrott H (2018) *Research at the Frontier of Knowledge: Comparing Text Similarity Indicators to Citations for Measuring Scientific Excellence*. Report TUM School of Management, Technische Universität München, Germany

Grant J, Brutscher P-C, Kirk S, Butler L & Wooding S (2010) *Capturing Research Impacts: A Review of International Practice*. Cambridge, UK: Rand Europe

Hicks D, Wouters P, Waltman L, de Rijcke S and Rafols I (2015) The Leiden Manifesto for Research Metrics. *Nature* 520: 429–431

Laudel G and Gläser J (2014) Beyond breakthrough research: Epistemic properties of research and their consequences for research funding. *Research Policy* 43: 1204–1216

Lebel J and McLean R (2018) A better measure of research from the Global South. *Nature* 559: 23–26

McLean R and Sen K (2018) Making a difference in the real world? A meta-analysis of the quality of use-oriented research using the Research Quality Plus approach. *Research Evaluation*: online. https://doi.org/10.1093/reseval/rvy026

Merton R (1973) Recognition and excellence: Instructive ambiguities. In: R Merton (ed.) *The Sociology of Science*. Chicago: University of Chicago Press. pp. 419–438

Moher D, Naudet F, Cristea I, Miedema F, Ioannidis J and Goodman S (2018) Assessing scientists for hiring, promotion, and tenure. *PLoS Biology*. doi:10.1371

Moore S, Neylon C, Eve M, O'Donnell D and Pattinson D (2017) 'Excellence R Us': University research and the fetishisation of excellence. *Palgrave Communications* 3 (January): 16105

Ndofirepi A (2017) African universities on a global ranking scale: Legitimation of knowledge hierarchies? *South African Journal of Higher Education* 31(1). doi: 10.20853/31-1-1071

OECD (1997) *The Evaluation of Scientific Research: Selected Experiences*. Paris: Organisation for Economic Co-operation and Development

Ofir Z and Schwandt T (2012) *Understanding Research Excellence at IDRC: Final Report*. Corporate Strategy and Evaluation Division, Ottawa, Canada: IDRC

Ofir Z, Schwandt T, Duggan C and McLean R (2016) *Research Quality Plus: A Holistic Approach to Evaluating Research*. Ottawa, Canada: IDRC

Schmidt E and Graversen E (2018) Persistent factors facilitating excellence in research environments. *Higher Education* 75(2): 341–363

Sørensen M, Bloch C and Young M (2015) Excellence in the knowledge-based economy: From scientific to research excellence. *European Journal of Higher Education* 6: 217–236

Tijssen R (2003) Scoreboards of research excellence. *Research Evaluation*: 12: 91–104

Tijssen R (2015) Research output and international research cooperation in African flagship universities. In: N Cloete, P Maassen and T Bailey (eds) *Knowledge Production: Contradictory Functions in African Higher Education*. Cape Town: African Minds. pp. 61-74

Tijssen R and Kraemer-Mbula E (2017) Perspectives on research excellence in the Global South: Assessment, monitoring and evaluation in developing country contexts. *SGCI Policy Brief* No. 1. December. https://sgciafrica.org/en-za/resources/Resources/SGCI%20Research%20Excellence%20Discussion%20Paper.pdf

Tijssen R and Kraemer-Mbula E (2018) Research excellence in Africa: Policies, perceptions and performance. *Science and Public Policy* 45: 392–403

Tijssen R, Visser M and Van Leeuwen T (2002) Benchmarking international scientific excellence: Are highly cited research papers an appropriate frame of reference? *Scientometrics* 54: 381–397

Tijssen R and Winnink J (2018) Research excellence in the Global South: Bibliometric evidence of 21st century trends. *CWTS blog post*, 2 October. https://www.cwts.nl/blog?article=n-r2u294&title=research-excellence-in-the-global-south-bibliometric-evidence-of-21st-century-trends

Young M (2015) Shifting policy narratives in Horizon 2020. *Journal of Contemporary European Research* 11(1): 16–30

CHAPTER 5

Gender diversity and the transformation of research excellence

Erika Kraemer-Mbula

Introduction

Female scientists and researchers play an essential role in contributing to development and transformative change. Gender equality, sustainability and development are highly interconnected (Leach et al. 2015). In fact, it has been argued that achieving the Sustainable Development Goals (SDGs) inescapably requires considering a gender dimension in research (Waldman et al. 2018).

The gender scientific gap has narrowed over the last decades, and women have had significant gains in terms of university enrolments worldwide. However, despite recent progress, the gender gap appears to persist, as women continue to experience numerous disadvantages that manifest in their academic careers: they are promoted more slowly than men, remain persistently under-represented in leadership research positions and agenda-setting roles, earn less than their male counterparts, tend to receive lower amounts of research funding, publish significantly less and are less cited, to mention a few examples.

Many studies have demonstrated the value of diversity in any type of organisation. For instance, studies on the business community have shown that having more female board members in firms has a positive effect on sales and returns of invested capital (Hunt et al. 2015), and firms with higher gender diversity display higher levels of innovation (Garba and Kraemer-Mbula 2018).

Greater diversity leads to better collective performance; this applies to research too. However, certain fields of science continue to have a strikingly low participation of women – for instance in engineering, physics and computer science there is less than 30% participation in most countries, with declining figures (WISAT 2012). Persistent gender imbalances in science, both in the Global South and globally, as well as insufficient progress in gender equality raise important questions for research excellence.

Gender disparities in research performance

Although there are more female than male undergraduate and graduate students in many countries around the world, women still represent a small percentage of researchers worldwide. The UNESCO Science Report (2015) indicates that women account for 53% of the world's bachelors and masters graduates and 43% of PhDs, but they only constitute 28% of researchers. Women also remain vastly under-represented at senior levels in scientific institutions. There are relatively few female full professors, and gender inequalities persist in hiring, earnings, funding and patenting (Lariviere et al. 2013).

Meritocracy in connection with research excellence builds on the basis that researchers should be rewarded on an objective basis, using clear and quantifiable criteria that enable distinguishing outstanding researchers from the average. Such 'objective' parameters commonly used to measure research excellence are based on quantitative indicators (mostly number of publications and citations). However, meritocracy applied as the sole basis to measure excellence seems to contribute to the reproduction of gender inequalities in academia.

Numerous large-scale studies continue to show that men publish more papers on average than women (Larivière et al. 2013; West et al. 2013; Bendels et al. 2018). Over and above total numbers, female authors are far less likely to publish single-authored papers, and in co-authored publications they are much less likely to be listed in a key position in a paper (usually considered as first author) (Bendels et al. 2018). Women are also less likely to publish in top-rated journals; this applies to all disciplines. Many studies across various disciplines confirm that female authors attract fewer citations than their male counterparts, and this applies also to high-impact science papers. Moreover, studies by Larivière et al (2013) and Bendels et al (2018) show that papers with female authors in key positions are cited less than those with male authors in key positions.

So what explains these differences in research performance? There is no consensus on the reasons for these gender differences in research outputs; however, the literature provides a range of explanations.

One of the underlying reasons often mentioned relates to widely held social stereotypes of gender and science. There is a general tendency to associate men with science and career, and women with liberal arts and family. Large-scale studies have found that 70% of men and women across 34 countries view science as more male than female (Nosek et al. 2009). It is difficult to assess how these social stereotypes may shape decision-making in various aspects of the research activity, from career choices among females to assessments of competence when hiring and promoting researchers.

Related to this argument is the difference across disciplines, whereby in terms of career preferences, women are conventionally associated with a preference towards careers focused on people, which would manifest an inclination for social sciences and humanities. Moreover, natural sciences, engineering, technology and mathematics are not typically portrayed as career-appropriate choices for women (Dugan et al. 2013). While studies indicate that there is a higher presence of female authors in disciplines in the social sciences, humanities is still dominated by men (Larivière et al. 2013).

Another explanation has to do with women's life-cycles, family, maternity and child care. This argument builds on the overlap between the critical years of research performance and women's fertility years, which leaves many women with the choice of either bearing children or gaining tenure (Jacobs and Winslow 2004; Ceci and Williams 2010).

This argument has been associated with the higher rate of withdrawal of women from scientific careers (Ceci and Williams 2011) or the tendency of women scientists to choose to work in lower-ranked universities, or end up in part-time, seasonal academic jobs, or administrative roles in universities (Wolfinger et al. 2009).

Women often interrupt their research careers due to childbirth and these gaps are often not taken into account in considerations for tenure. In this respect, a study by Hunter and Leahey (2010) calculates the effect of childbirth on publications, estimating it at two years of lost publications. The effect of having young children (under ten years old) on the productivity of men and women has not been clearly established. However, it is known that women tend to acquire most of the caregiving responsibilities in the early years of childcare.

Ceci and Williams (2011) argue that the critical variable that explains the lower research performance of women may not be related to gender directly, but to access to resources which correlates with gender since women are more likely to work in positions or institutions with limited resources. In certain fields of science, women generally lead smaller labs and draw fewer resources, leading to fewer opportunities for career advancement (Murray and Graham 2007).

An important aspect of excellence has to do with recognition by peer scientists. In this respect, it has also been argued that women are less integrated in professional networks than men. Academic institutions have predominantly male professional cultures, which often make female scientists feel isolated and excluded from social circles in science where resources, knowledge and reputation are exchanged and developed (Etzkowitz et al. 2000). Having lower levels of social capital also translates into less participation in international research projects, less international collaborative publications, and less citations. Krefting (2003), in a study of USA universities, explains that while women and other minorities have entered universities, they are still outsiders to the academic game. In this respect these groups may find it relatively harder to make sense of the organisational structures and of the values of the universities that employ them.

challenges in the careers of African female scientists relate to lack of mentoring and lack of mobility and training opportunities.

Adding to these factors, the internationalisation of academic careers in the Global South also plays an important part in creating a context of intense competition, with the promotion of 'excellence' as the central criterion in academic promotions, particularly in professorial ranks. Although the number of females eligible to apply for promotions has increased considerably due to the growth in the participation of women in higher education, gender disparities persist in the scientific workforce. Female scientists remain concentrated in posts with lower responsibility and decision-making and limited leadership opportunities. For instance, data from the Higher Education Management Information System in South Africa in 2016 show that 58% of higher education students were women. However, there is a drop in the number of women along the career trajectory in scientific research. While at junior lecturer and lecturer levels, women make up 53% of total posts, at senior lecturer level the number decreases to 45%, and only 27% of professors in South African institutions are female. In Cameroon, enrolment in tertiary education was estimated at 15% for women in 2017, while women constituted only 7% of academics at the rank of full professor (UNESCO 2018). As expressed by Huyer (2015: 86): 'Each step up the ladder of the scientific research system sees a drop in female participation until, at the highest echelons of scientific research and decision-making, there are very few women left'.

However, the under-representation of women in research and leadership positions in universities in the Global South is, at the same time, subject to other (global) imbalances. In this respect, it is important to be reminded that gender is deeply interwoven with other dimensions that shape power relations in research activities and processes, such as race, class, ability, sexuality, location, etc. (Cornwall and Sardenberg 2014). Therefore, considerations of research excellence cannot be seen as separate from broader geopolitical forms of dominance, in which perspectives of Southern researchers remain marginalised. In exploring the inclusion of scholars in the South in global knowledge production, a recent study by Medie and Kang (2018) analysed the institutional affiliation of authors published in journals related to women, gender and politics and found that South-based scholars constituted less than 3% of the articles in four leading European and North American journals between 2008 and 2017. The authors argue that such under-representation of scholars in the Global South 'demonstrates the hegemony of Western gender politics scholarship and reinforces the power disparity in knowledge production between the North and South' (Medie and Kang 2018: 38).

Considerations of gender equality in research excellence in the Global South must therefore address unequal power relations on a range of social and political dimensions at multiple scales from the personal to the global.

Moving towards diversity thinking in research excellence

This chapter has identified the various dimensions where gender bias can be identified in relation to academic performance and research excellence. In addition to gender bias, systematic constraints built into academic institutions have played an important role in impeding the careers of women scientists throughout modern history.

Some authors, inspired by practices in large private firms, have proposed a framework that incorporates three phases in the evolution of diversity, from 1.0 to 2.0 and 3.0 (Nivet 2011; Sepulveda et al. 2018).

The goal for the Diversity 1.0 phase is to alleviate institutionalised discrimination to seek fairness and equality with respect to gender and ethnic differences. The actions under Diversity 1.0 tend to be isolated efforts and programmes aimed at removing social and legal barriers to access and equality. Diversity 2.0 actions are often geared towards raised awareness about how increasing diversity benefits everyone, expanding the programmes initiated under Diversity 1.0, but still keeping diversity on the periphery rather than becoming part of the core mission of institutions. The next paradigm, Diversity 3.0, is fueled by the understanding that diversity and excellence are not only complementary, but also intricately linked. Under Diversity 3.0, diversity and inclusion become central to the institutional mission and integral for achieving excellence.

Table 1: Phases in the evolution of diversity thinking

Diversity 1.0	Isolated efforts aimed at removing social and legal barriers to access and equality, with institutional excellence and diversity as competing ends.
Diversity 2.0	Diversity kept on the periphery but raised awareness about how increasing diversity benefits everyone, allowing excellence and diversity to exist as parallel ends.
Diversity 3.0	Diversity and inclusion integrated into the core workings of the institution and framed as integral for achieving excellence.

Source: Nivet (2011)

In line with Diversity 1.0, universities in the Global South have generally developed anti-discrimination laws to remedy conditions that differentially affect women's entry into and promotion in academic scientific and research careers. These laws often accompany broader national and regional recognition of the importance of women's right to development, such as the 2015 declaration by the African Union as the 'Year of Women's Empowerment and Development'. However, the existence of institutional and legislative frameworks, designed to transform academic institutions around the principles of non-sexism and non-racialism, does not always translate into the realisation of equality. There is a lack of mechanisms to enforce anti-discriminatory legal frameworks, for example monitoring and evaluation systems. For instance, while legal trends recognise that stereotyping is a form of discrimination, the extent to which stereotyping practices continue to limit women's advancement remains largely undocumented. Other steps needed to remove barriers include documenting the status and progress of under-represented groups and establishing a work environment that is explicitly inclusive.

Second-generation gender bias can be found under Diversity 2.0, where legal frameworks may exist at the institutional level, and even in isolated programmes that promote inclusion and equality; however, subtle barriers for the advancement of women persist, including cultural assumptions, organisational structures, and practices and patterns of interaction that inadvertently benefit men. For instance, when more men are in leadership positions in a research environment, this can potentially result in weaker networks for women. By supporting male-led networks at the top, even without discriminatory intent, such practices can obstruct leadership in women scientists. Another example is the current model of a scientific career, which is still built on an outdated model of a male life course. Under prevailing career models, researchers in high positions are expected to have unlimited commitment to their academic careers throughout their working life. This model, which depends on having a spouse who takes care of the household, family and community, is increasingly unfitting to not only most women but also men. These examples serve to illustrate that under Diversity 2.0, education and awareness actions about the collective benefits of gender diversity and equality coexist, with seemingly 'neutral' approaches towards valuing and supporting excellence that continue to limit the advancement of women.

Recognising the importance of supporting women scientists, universities, research funders and scientific and professional associations have developed a range of programmes and mechanisms designed to assist women scientists at the early stages of their career, as well as those already in posts, often providing mentorship and training. However, these efforts often remain isolated efforts and are not fully embedded in institutional practices.

Under Diversity 3.0, diversity and inclusion would be integrated into the institution's core functions and into the framework for achieving excellence. Judging by the results, this is far from a reality in the research environment in the Global South. Gender diversity remains a challenge for academic and research organisations not only in the Global South, but globally. In order to achieve the broad aspirations of diversity, equality and empowerment, diversity must become integral to achieving excellence.

Some important initiatives have been recently captured in a report led by the Gender Working Group at the Global Research Council (GRC 2019), showing the efforts that research councils around the world are making towards promoting a research environment which more fully supports the equality and status of women in research.

Creating a research culture that is respectful, diverse and inclusive fosters academic excellence and broadens perspectives. Our current global challenges are daunting and demand multifaceted knowledge.

Besides the moral imperative of embracing diversity and inclusion, fostering diversity in institutions such as research councils and universities adds to building inclusive systems and enhances systemic creativity, innovation and problem-solving.

Concluding remarks

The academic research environment is characterised by the under-representation of women, persistence of a masculine culture and the model of an excellent scientist reflects an outdated male life-cycle, restricting recognition of work done outside academia. Therefore, looking at excellence from the lens of conventional 'neutral' indicators continues to suggest that research excellence is largely a male territory. Excellence may not be intentionally a masculine construct but its application in the academic system is. In connection to this argument, a study by Feller (2004) explores the difference between bias present in the system and bias present in the indicator. It is therefore important to both question the indicators used to measure excellence – perhaps thinking around measurements of 'collective excellence', as well as explore the persistent exclusion mechanisms for women in the academic system.

In addition to the global imbalances, women in the Global South experience specific challenges that relate to the context in which they operate. It is thus important to explore gender as one of the several dimensions that shape power relations in academic environments in the Global South. These aspects continue to receive little attention and need to be unpacked.

Finally, although there is a long way to go, there are ongoing efforts that research councils, research funders and research organisations are already undertaking in order to move towards a more equitable future in scientific research. This chapter presents diversity 3.0, namely the integration of diversity into institutional frameworks in order to achieve excellence, as an essential step forward in building inclusive research systems.

References

Arocena R, Göransson B and Sutz J (2018) *Developmental Universities in Inclusive Innovation Systems: Alternatives for Knowledge Democratization in the Global South*. London: Palgrave Macmillan

Bendels MH, Müller R, Brueggmann D and Groneberg DA (2018) Gender disparities in high-quality research revealed by Nature Index journals. *PloS One* 13(1): e0189136

Bezuidenhout A and Cilliers FV (2010) Burnout, work engagement and sense of coherence in female academics in higher-education institutions in South Africa. *SA Journal of Industrial Psychology* 36(1): 1–10

Ceci SJ and Williams WM (2011) Understanding current causes of women's underrepresentation in science. *Proceedings of the National Academy of Sciences* 108(8): 3157–3162

Ceci SJ and Williams WM (2010) Sex differences in math-intensive fields. *Current Directions in Psychological Science* 19(5): 275–279

Cornwall A and Sardenberg C (2014) Participatory pathways: Researching women's empowerment in Salvador, Brazil. *Women's Studies International Forum* 45: 72–80

Dugan JP, Fath KQ, Howes SD, Lavelle KR and Polanin JR (2013) Developing the leadership capacity and leader efficacy of college women in science, technology, engineering, and math fields. *Journal of Leadership Studies* 7(3): 6–23

Etzkowitz H, Kemelgor C and Uzi B (2000) *Athena Unbound: The Advancement of Women in Science and Technology*. Cambridge: Cambridge University Press

Feller I (2004) Measurement of scientific performance and gender bias. In: E Addis and M Brouns (eds) *Gender and Excellence in the Making*. Brussels: Directorate General for Research, Science and Society Series

Garba T and Kraemer-Mbula E (2018) Gender diversity and enterprise innovative capability: The mediating effect of women's years of education in Nigeria. *International Journal of Gender and Entrepreneurship* 10(4): 290–309

Global Research Council (GRC) (2019) *Supporting Women in Research: Policies, Programs and Initiatives Undertaken by Public Research Funding Agencies*. Report by the The Gender Working Group (GWG). https://www.globalresearchcouncil.org/fileadmin/user_upload/GRC_GWG_Case_studies_final.pdf

Gwele NS (1998) Gender and race: perceptions of academic staff in selected faculties in English language historically white universities concerning their working conditions. *South African Journal of Higher Education* 12(2): 69–78

Hunt V, Layton D and Prince S (2015) *Why Diversity Matters*. McKinsey.

Hunter LA and Leahey E (2010) Parenting and research productivity: New evidence and methods. *Social Studies of Science* 40(3): 433–451

Huyer S (2015) Is the gender gap narrowing in science and engineering? *UNESCO Science Report: Towards 2030*. Paris: UNESCO

Jacobs JA and Winslow SE (2004) The academic life course, time pressures and gender inequality. *Community, Work & Family* 7(2): 143–161

Joubert M and Guenther L (2017) In the footsteps of Einstein, Sagan and Barnard: Identifying South Africa's most visible scientists. *South African Journal of Science* 113(11-12): 1-9

Kraemer-Mbula E (2014) University linkage and engagement with knowledge users at community level. In: GTG Mohamedbhai, G Frempong and A Addy (eds) *University Research Governance & National Innovation Systems in West and Central Africa*. Accra: Association of African Universities. pp.169–189

Krefting LA (2003) Intertwined discourses of merit and gender: Evidence from academic employment in the USA. *Gender, Work and Organization* 10(2): 260–278

Larivière V, Ni C, Gingras Y, Cronin B and Sugimoto CR (2013) Bibliometrics: Global gender disparities in science. *Nature News* 504(7479): 211–213

Leach M, Mehta L and Prabhakaran P (2015) *Gender Equality and Sustainable Development.* London: Routledge

Maslach C and Leiter MP (1997) *The Truth about Burnout: How Organisations Cause Personal Stress and What to do About it.* San Francisco, CA: Jossey-Bass

Medie PA and Kang AJ (2018) Power, knowledge and the politics of gender in the Global South. *European Journal of Politics and Gender* 1(1-2): 37–53

Mohamedbhai GTG, Frempong G and Addy A (eds) (2014) *University Research Governance & National Innovation Systems in West and Central Africa.* Accra: Association of African Universities

Murray F and Graham L (2007) Buying science and selling science: gender differences in the market for commercial science. *Industrial and Corporate Change* 16(4): 657–689

Nivet MA (2011) Commentary: Diversity 3.0: A necessary systems upgrade. *Academic Medicine* 86(12): 1487–1489

Nosek BA, Smyth FL, Sriram N, Lindner NM, Devos T, Ayala A et al. (2009) National differences in gender–science stereotypes predict national sex differences in science and math achievement. *Proceedings of the National Academy of Sciences* 106(26): 10593–10597

Prozesky H and Mouton J (2019) A gender perspective on career challenges experienced by African scientists. *South African Journal of Science* 115(3-4): 1–5

Rothmann S and Barkhuizen N (2008) Burnout of academic staff in South African higher education institutions. *South African Journal of Higher Education* 22(2): 439–456

Sepulveda KA, Paladin AM and Rawson JV (2018) Gender diversity in academic radiology departments: Barriers and best practices to optimizing inclusion and developing women leaders. *Academic Radiology* 25(5): 556–560

Tijssen R and Kraemer-Mbula E (2018) Research excellence in Africa: Policies, perceptions, and performance. *Science and Public Policy* 45(3): 392–403

UNESCO (United Nations Educational, Scientific and Cultural Organization) (2015) *UNESCO Science Report: Towards 2030.* UNESCO

UNESCO (2018) Institute of Statistics. http://uis.unesco.org/

Waldman L, Abreu A, Faith B, Hrynick T, Sánchez de Madariaga I and Spini L (2018) *Pathways to Success: Bringing a Gender Lens to the Scientific Leadership of Global Challenges.* Institute of Development Studies

West JD, Jacquet J, King MM, Correll SJ and Bergstrom CT (2013) The role of gender in scholarly authorship. *PloS One* 8(7): e66212. https://doi.org/10.1371/journal.pone.0066212

WISAT (2012) National Assessments on Gender Equality in the Knowledge Society: Gender in Science, Technology and Innovation. http://wisat.org/wp-content/uploads/GEKS_Synthesis-Nov2012.pdf

Wolfinger NH, Mason MA and Goulden M (2009) Stay in the game: Gender, family formation and alternative trajectories in the academic life course. *Social Forces* 87: 1591–1621

CHAPTER 6

Research excellence is a neo-colonial agenda (and what might be done about it)

Cameron Neylon

Introduction

The pursuit of 'excellence' is central to the identity of today's researchers, research institutions, funders and national research strategies. Most funders and national policies reference 'excellence' or 'quality' as one of the main criteria for support. Researchers advocate for the importance of their own work with claims of 'excellence', bringing a wide range of evidence to support their arguments. However, it is rare for these terms to be clearly defined or for common definitions to be agreed on.

Even where policy agendas seek to support qualities of research that lead to outcomes, engagement or wider impacts, care is taken to distinguish between traditional conceptions of research excellence, and these 'new', 'complementary', or 'expanded' aspects of evaluation (Donovan 2007). Researchers in turn seek to reinforce this dichotomy by claiming that agendas for impact and engagement risk damaging research excellence (Chubb and Watermeyer 2017). The argument that there is little distinction in practice between outcomes and the impact on further scholarship and outcomes and impacts that occur in the wider community[1] have largely been ignored in favour of a sharp distinction between 'excellence' and 'impact' (Donovan 2007).

Yet, as we have previously argued (Moore et al. 2017), this concept of 'excellence' is an empty rhetorical construct with no common meaning and no value. In fact, it is deeply damaging to the production of research with relevance and importance to actual policy goals, development and the improvement of wider publics, as well as to the qualities of curiosity-driven research it is supposed to protect. It drives instrumental, rather than values-based and normative behaviour and is at the centre of almost every problem facing the Western academy, from issues of diversity, inclusion and bias, to the rise in fraud and malpractice.

All of these issues are further compounded in the context of countries that are outside the traditional power centres of Western scholarship. Control of the systems of research communication, and current modes of evaluation, is firmly vested in the hands of North American and European scholarly institutions and corporations. The historic development – both positive and negative – of our conceptions of the proxies and signals of research excellence is entirely based on the concerns of countries close to the North Atlantic,[2] with an equally narrow literature, modes of assessment and service providers.

The form and structure of research institutions in many countries, particularly south of the equator, is a product of colonial and post-colonial histories. For example, in South Africa most of the older institutions of higher education and research have explicitly British or Afrikaner origins. Institutions founded after independence have their own character and challenges rooted in the particular historical issues of South Africa and in the apartheid and post-apartheid period (Soudien 2015). All of South Africa's institutions are grappling with the question of decolonisation and its challenges (Joseph Mbembe 2016). Many of these challenges are common to other post-colonial countries.

In this piece I want to argue that, while the agenda for research excellence is connected strongly to this colonial and post-colonial history, the agenda is in fact *neo-colonial*. Recent work shows that our current conceptions of research excellence and their signals only arose over the past 50 years. This suggests that their adoption and spread through countries with a colonial legacy should not be seen only as a consequence of history, but also as a new wave of epistemic colonisation. This distinction offers important ways to recognise, tackle and address the problems and opportunities in a post-colonial context and suggests ways in which these countries can provide leadership to and build community with other post-colonial, developing and transitional nations. More than this, it can help us to understand how these experiences can provide leadership to Europe, North America and other traditional centres of Western scholarship that appears unlikely to arise internally.

A brief speculative history of research excellence

One of the challenges in this space is that historical analysis of post-1945 development of research institutions and culture is both sparse and challenging. What follows is therefore of necessarity a speculative and anecdotal description, rather than a rigorous historical analysis.[3] This is an important area for future research.

Prior to 1945, research and scholarship was largely the preserve of clubbish institutions in the countries and regions bordering the North Atlantic. Arguments about what constituted 'good work' or a 'good scholar' have a long history. The broad form of these arguments was largely focused on who would be allowed into those traditional clubs with national academies, such as the Royal Society of the United Kingdom (UK), being a significant focus.

After 1945 there was a massive expansion of national funding of research, firstly in Europe and North America, but later globally. Universities in colonial settings including Africa and Latin America, but also countries such as Australia, which had been largely built for the local training of professional classes, or for education of the children of colonial administrators, grew as research centres in their own right, and then as centres of national pride and prestige with independence.

This expansion of both the scale of research and number of researchers and of state investment with its consequent focus on the productivity of that investment led to a range of challenges for the academy. First, the club-based modes of evaluation in which personal recommendation and direct knowledge of the researcher being evaluated broke down as the size of the community grew. Simultaneously the growth of government interest in the deployment of their investment led to deep anxiety about the autonomy of research institutions.

As Baldwin (2017) and others have noted, it is these two strands that led to the institutionalisation of peer review. Peer review functioned both as a means of establishing autonomy of the academy – only peers can do peer review – and through the standardisation of the process of review, which allowed the scholarly literature to scale up, while still having its boundaries clearly defined. The scaling up of the journal literature meant that it was necessary to develop common protocols that defined what would count as 'scholarly'. Peer review came to serve that function, but it was only from the 1970s on that it was considered a universally necessary component of scholarly publishing.

Later, the 'impact agenda' grew out of a similar concern for governments' and funders' interest in understanding and maximising the economic impact of research. In the UK and Australia particularly, research communities mobilised against this narrow scope of assessment and the idea of 'wider impacts' was developed, particularly in Australia (Donovan 2008). Broadly speaking, the research community remains opposed to these agendas, as they threaten the autonomy of the academy to set its own priorities, and makes academic work subordinate to the needs of the community or the state (Smith et al. 2011).

'Research excellence' is often deployed in dichotomous opposition to impact and societal engagement agendas as a way of defending autonomy. For instance, in the work of Chubb and co-workers (Chubb and Watermeyer 2017; Chubb and Reed 2018) based on interviews with researchers in Australia and the UK on their experience of requirements for grant submissions, interviewees objected to the way in which impact requirements lead them to overstate claims or indeed lie. This is implicitly contrasted with the serious and rigorous approach which the interviewees claim is applied to the description of the research outcomes themselves.

This deployment of research excellence as a rhetorical tactic to defend autonomy has many parallels with the development of peer review 40 years earlier. It arrogates assessment to internal mechanisms of the academy, and it privileges the standing of traditional centres of power and senior leadership to describe, evaluate and embody that excellence. While the tactics have been largely successful, the increasing scrutiny of governments has required that the academy present more substantial evidence of this claimed research excellence. Simple claims of expertise and authority are no longer sufficient. This in turn has led to a heavy reliance on supposedly objective measures such as citation-based proxies.

Not surprisingly this has coincided with an increase in the availability and use of citations as a proxy or correlate of 'excellent' research. The availability of data through the release of Science Citation Reports led to debate on the meaning of the data, which ultimately gave rise to the assumption that citations were a measure of 'research impact' borrowing from the term 'Impact Factor', coined by the Institute for Scientific Information (see Bornmann and Daniel 2008 for a review of this debate).

The assumptions that such quantitative data are in any sense objective, that they represent appropriate incentives for the research community, or that quantitative assessment and rankings of any sort are appropriate, have come under significant criticism since they were introduced. Nonetheless, concepts such as the primacy of citations, the importance of journal brand and impact factors, H-indices and institutional rankings have rapidly become deeply embedded in the assumptions and practice of the academy globally.

Research excellence as a neo-colonial agenda

The challenge of confidence and quality

Many of the challenges facing countries seeking to develop their research capacity can be seen through the lens of self-confidence. When compounded with resource limitations this leads to a perceived need for external validation and certification.[4] A concern for effective investment requires identifying research and researchers of high quality that justify the investment being made. In turn, this leads to a search for 'objective' and 'international' measures that can be used to determine quality. In contexts with a history of corruption or nepotism, the perceived need for outside objective validation can be very strong.

This lack of confidence, both as individual decision-makers, and more broadly in the sense of subjugation vis à vis the North Atlantic, is in many cases a colonial legacy. The systematic disruption of indigenous and local systems of knowledge, governance and communication and their replacement with those of the controlling power was a core part of the colonial system. Similarly, the legacy research institutions and the global system of research communication are explicitly colonial systems.

Building a new academy founded on local needs and values which also interfaces with the international system is difficult. Rebuilding locally founded capacity and confidence, while also having the internal capacity to identify what is valuable in the 'international' system can be – or at least can be perceived to be – at odds. In particular, there is a risk of the same false dichotomy discussed above being set up. In other words, the setting of local priorities towards societal engagement and wider impacts is positioned as being in opposition to 'objective' and 'international' measures of 'excellence'.

In addition, those who were brought up and achieved success in colonial and post-colonial systems, whether locally or in the institutions of colonial powers, are invested in that particular form of autonomy for the academy which is aligned with European and North American (North Atlantic) ideas of excellence. Autonomy of scholarship is critical for a developing or transitional country. It is an important part of building productive institutional forms for a pragmatic and modern knowledge-based state. A well-functioning academy will balance a necessary separation from the state to preserve its autonomy and freedom to examine, criticise and recommend, while sharing the concerns of the state, and of various communities, to deliver scholarship for the public good.

There are serious difficulties in simultaneously building confidence in local capacity and expertise, gaining sufficient confidence of government and the state to build institutional autonomy, and developing a strong culture of *internal* assessment that builds on strengthened culture and values.

The neo-colonial nature of available proxies

In the context of this struggle for decolonisation, the appeal of reaching for 'international' and 'objective' measures for validation is obvious. Numbers offer the illusion of these qualities, but in fact the numbers available do not deliver them (Wouters 2016). They are not objective in as much as they are based on opaque and commercially focused selection decisions. They are not international, because they are built almost exclusively on the historical needs of North Atlantic American researchers, publications venues and publishers.

Once more, the agenda of Europe and North America dominates the discourse, describing what matters and what is important. That which is considered important in Cambridge, for example, is 'international', whereas that which is important in Ubatuba, Hanoi or Lagos is merely 'local'. These surface issues are well discussed. What is more problematic is the much deeper integration of this 'international' system of scholarship into organisations running to European and North American imperatives. Just as the two East India companies, running from Amsterdam and London, sought to control the modes, mechanisms and infrastructures of trade in the 17th and 18th centuries, multinationals based out of those same cities dominate the infrastructures of research assessment and communication.

Just as the expansion of international trade was driven by a gradual depletion of accessible natural resources in Europe and North America and the massive opportunities that new transport technologies brought to exploit resources in Africa, South America and South East Asia, companies today are seeking new resources. With a limited scope for increasing market size and revenue in the saturated markets of the North Atlantic Region, the web enables Clarivate and Elsevier (as well as other companies and non-profits) to pivot to a new set of countries, including the post-colonial nations,[5] investing in the expansion of their knowledge base and institutions as new markets to grow.[6]

This is therefore a process of re-colonisation. If 'data is the new oil', then expropriation of data, knowledge and human capacity by powerful corporate and state actors is a logical consequence. As with the colonisations of the 17th to 19th centuries, this starts by imposing the governance and systems of the colonising powers. Technical infrastructures, forms of evaluation and the data that support them are all controlled by powerful corporate actors, with no significant oversight of their governance, selection processes or design.

As with previous cycles of colonisation, these systems were built largely for North Atlantic customers to benefit largely North Atlantic investors and then provided to the rest of the world with the claim that they are 'neutral', 'objective' and 'international'. As with previous cycles, the interlocking institutions of evaluation, resourcing, recording and dispute resolution are coupled together to make it difficult to engage with just a part of the system and close to impossible to unpick the pieces once they are implemented. In this sense, the East India companies were early masters of vertical integration as a business strategy.

The Sabato Triangle in a networked world

Just over 50 years ago, Sabato and Botana (1968) released a paper that has apparently never been translated into English (see also Chapter 2 by Sutz in this volume for more details). First presented at the World Order Models Conference and published in *Revista de la Integración*, the paper *La Ciencia Y La Tecnología En El Desarrollo Futuro De América Latina* provides a model of how different sectors combine to support development within a nation. Some 30 years before Etzkowitz and Leydesdorff (1995) proposed the Triple Helix Model, Sabato and Botana described how government, industry and knowledge production sectors needed to interact and build on each other to deliver development. This is represented as a triangle, with the corners representing each sector (see Figure 1).

Central to Sabato and Botana's argument is that, for development, the strength of each corner is less important than a balance of the interactions between them. In particular, they point out that a specific failure mode arises when one of the corners has stronger interactions with the 'international' system than with the other sectors of the local system of development. In their view, the failure of earlier programmes of development that combined parallel investments in industrial

Figure 1: The Sabato-Botana Triangle. Adapted from Sabato and Botana (1968). 'Gobierno' is the system of government, 'Estructura productiva' is the industrial system and 'Infraestructura cientifico-tecnológica' is the scientific/technological system of research.

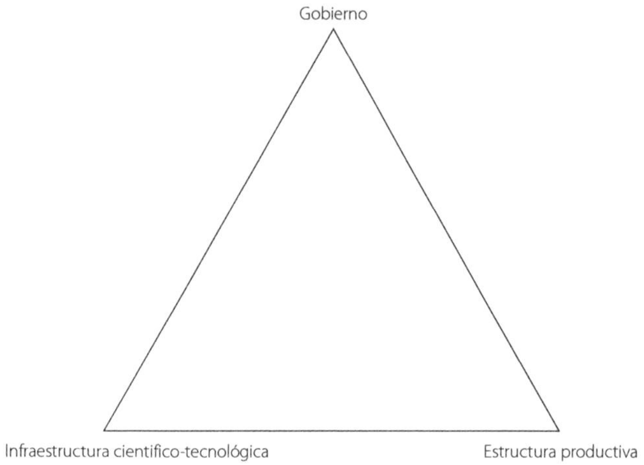

capacity with investments in knowledge production and technology was being caused by a lack of interaction between the sectors that are intended to arise from these investments.

The modern objections to the Sabato-Botana Triangle model are that it is too simplistic and creates too inflexible a relationship between the three sectors. As with the Triple Helix, we might also argue for the addition of a fourth corner, civil society and the media, as more fully reflecting the interconnections in society (Carayannis and Campbell 2009). Nonetheless, the Triangle as a conceptual model offers a valuable way to complement classical analyses such as Dependency Theory and Decolonisation in providing a framework that emphasises the importance of the *interconnectedness of the local* alongside the importance of valuing the local.

To apply the Sabato-Botana model in a networked world (Figure 2), it is necessary to break down the more rigid categorisation implied by the sharp corners to consider agents, and their connections. This provides a powerful way of analysing how different actions and players

Figure 2: Adapting the Sabato-Botana triangle to a network view. The three vertices of the triangle represent well-interconnected groups within broader society. Some actors will bridge between groups and play an important role in creating and maintaining links. Some of these links can be tracked and monitored with available data, primarily through citation and co-authorship links within the scientific-technological system.

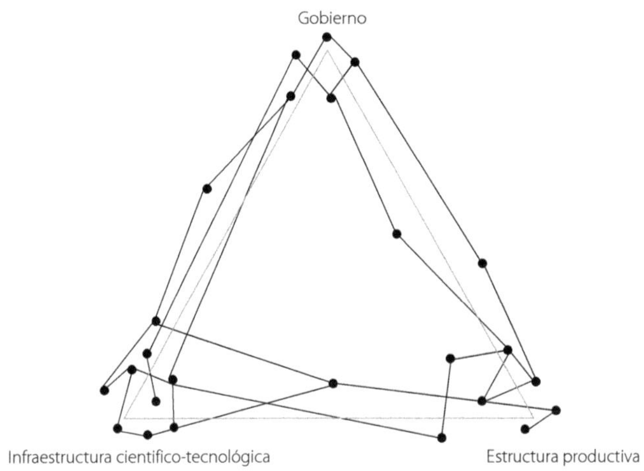

strengthen and weaken connections, either within the local triangle or outside it (see Figure 3). More than this, we can probe our ability to ask these questions and identify gaps in our knowledge that would help us to track the creation, breaking, strengthening and weakening of these connections.

Sabato and Botana note one form of this in the 1968 paper, describing the loss of talent to overseas systems:

> *En América Latina, el éxodo de talentos es la típica consecuencia de la falta de inter-relaciones entre la infraestructura científico–tecnológica, la estructura productiva y el gobierno. Por esta razón, los científicos formados en nuestras sociedades, faltos de incentivos, se relacionan con una infraestructura científico–tecnológica del exterior. Pero al actuar así, el científico que emigra*

Figure 3: The biasing effect of strong interactions with the international research system. Rhetorics of 'research excellence' privilege connections of the form shown as arrows from the national/regional system to international connections. This weakens local relationships, both within the scientific-technical system and more broadly in society, including the fourth vertex of civil society that is not present in the original triangle model.

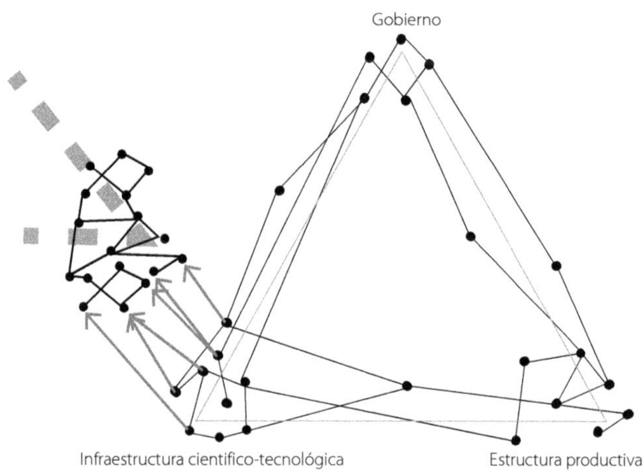

hacia los grandes centros de los países industriales, se integra en un triángulo de relaciones plenamente capacitado para satisfacer las demandas que plantea su tarea específica. Mientras en nuestras sociedades el científico se encuentra desvinculado y aislado frente al gobierno y a la estructura productiva, en el nuevo lugar de trabajo, al cual lo conduce su exilio cultural, está automáticamente amparado por instituciones o centros de investigación que, a su vez, se encuentran insertas en el sistema de relaciones que hemos explicado.

In Latin America, the loss of talent overseas is a typical consequence of the lack of connections between the scientific and technological structures, the industrial production structures, and the government. Scientists trained in our

> society are driven by incentives systems to align themselves with foreign scientific and technical infrastructures. Scientists who emigrate to the centres of scholarship in industrial countries become fully integrated into an existing triangle of relationships, fully equipped to meet the demands of their specific tasks. By contrast, in our societies scientists are disengaged and isolated from government and industrial structures. In the new workplace, to which their cultural exile leads them, they are protected automatically by institutions or research centres which are already engaged in these systems of relationships. [author's translation, based on Google Translate]

Today we can consider directly probing these processes. Do scholars who emigrate from post-colonial countries return? Do they strengthen local connections amongst scholars or simply strengthen the spokes of networks that have their hubs in the old colonial centres? Other chapters in this volume illustrate some of the ways this analysis can be tackled, alongside recent work by Sugimoto and co-workers (2017). More generally, we can examine the flow of citations, of the use of concepts and ideas, how this changes over time, and whether it is evidence of flows to those same traditional hubs, or of strengthening local connections and building local networks and hubs.

We can also critically examine what information we do not have that could aid in this analysis. There is a lack of information sources that would aid in tracking the strengthening and weakening of ties between the research, industrial and government systems in a consistent and scalable fashion. There is also a lack of coverage, even within the information on the research system, of journals based in post-colonial and developing countries, of languages other than English and of topics of interest beyond the North Atlantic.

We therefore have two interlinked questions. The first is which actions and choices strengthen the local ties that support development (and arguably innovation) in a balanced manner? The second question is what information gaps exist in seeking to answer the first question. The first question seeks to address issues that are frequently a colonial legacy. The second, and in particular the gaps being created by information gathering focused on narrow and North Atlantic-focused modes of evaluation and the bias towards measuring and valuing non-local connections, is neo-colonial.

This is true in terms of the immediate concern of how a post-colonial or transitional country is capable of evaluating its own progress, but the damage goes deeper than that. The North Atlantic focus of the data, combined with the narrow conception of 'research excellence' and corporate strategies of vertical integration that they are built around, deliberately undermine the ability of these countries to develop their own systems of strategic information. Again, this parallels the strategies of the corporate-state actors of the 18th century. The advantage that developing and transitional countries have is the ability to recognise that this is a *new* cycle of colonialism and to act accordingly with the knowledge of history.

How do we address these issues?

As noted above, and as is the case with decolonising agendas more generally, the question of how to respond is not straightforward. The challenge of capacity building in developing and transitional countries is a real one. A significant part of the colonial legacy was the weakening and destruction of local knowledge, communication and governance systems. Complete disengagement from colonial and neo-colonial systems is not an option. Nor, obviously, should complete acquiescence be. The challenge is in identifying which parts of these systems are valuable in a local context and how they might be appropriated. This is important because the model above, while it emphasises a focus on the formation and strengthening of local connections, does not provide an answer as to *which* connections will be of value in that local context.

Being internationally engaged is not inherently problematic. Building and strengthening local institutions of research and knowledge production that provide the capacity to appropriate and exploit knowledge produced in traditional North Atlantic centres of scholarship is a sensible policy goal. Building and strengthening a profile within the constraints of North Atlantic concepts of 'excellence' can also be a sensible tactical or strategic action in the context of building that capacity, attracting and retaining talent and investment. Appropriating and exploiting the affordances of platforms that support those systems may be a significant part of these approaches.

The challenge is in identifying which parts of the appropriated system are of local value, which will further structural bias, and how these are connected. The distinction Connell (2016) makes between 'Western' and 'imperial' knowledge may be of value here, provided we recognise the way in which the tools and approaches which may be of value in general (the 'Western') are tightly coupled to systems and processes which sustain the power imbalances that underpin the 'imperial' (Chan et al. 2018). Aspects of good practice articulated within agendas such as that for 'open scholarship' include reproducibility, transparency and effective communications. These may seem like unambiguously good approaches, but their implementation is also often tied to systems and structures that require access to significant – and costly – technical infrastructures, such as computational capacity and telecommunications networks (Chan et al. 2018).

Any such appropriation carries its own risks. These are 'the master's tools' after all (Lorde 1984). Lorde's call in the paper that starts with those words is to give space and voice to the disenfranchised. In this context it is critical to do more than merely listen, or merely incorporate those voices, but to create the institutional forms that *privilege* that diversity of voices. Lorde addresses this in the context of the necessity of a diversity of women's voices:

> Advocating the mere tolerance of difference between women is the grossest reformism. It is a total denial of the creative function of difference in our lives. Difference must be not merely tolerated, but seen as a fund of necessary polarities between which our creativity can spark like a dialectic.

The core problem of the rhetoric of research excellence is its homogeneity and its consequent privileging of North Atlantic and therefore inevitably white voices. It is this homogeneity, combined with existing structures of power and prestige, that is problematic. From university rankings to individual hiring decisions, this drives actors at all scales to seek to become the same. The mismatch between the apparent goal and the needs of society may be most obvious in post-colonial and transitional countries, but it is also a growing problem for scholars in the so-called first world. As we shall see, it is the experiences and culture of scholars and institutions beyond the traditional centres of prestige, whose creativity is already delivering, which may have more to provide than those at the traditional centre.

Shifting the narratives: The qualities of quality and privileging the interconnectedness of the local

Building a rigorous and contextualised framework for deciding this is beyond the scope of this paper. It requires a programme of political negotiation towards agreeing local needs and priorities, alongside a social model of knowledge creation that can manage the complex flows that include the special characteristics of local knowledge. More than this, it is inappropriate for an outsider, particularly one from 'the centre', to offer advice. Any such advice should be treated with suspicion. What I propose below should be seen as a tentative set of actions for local decision-makers to consider, critique and adapt as is needed.

Building institutions immediately raises one of the hardest problems to tackle, that of shifting culture and the narrative that supports it. This is long term and difficult work. However, careful choice of words and their deployment, or not, can be powerful. Here I want to tackle the use of two terms, 'excellence' itself and 'international' as examples of how deliberate choices in word usage can be helpful.

The first step is to reject rhetorical forms and narratives that support the idea of a unitary – and quantitative – concept of excellence. Often this seems obvious and easy. It is, however, extraordinarily radical.[7] It requires at its core the rejection of the idea that scholarship can be ranked. It may be *prioritised*, or *evaluated,* in a particular context of resource allocation, but an agenda of decolonisation requires that the idea that any given piece of scholarship can be objectively better than another be rejected.

While there is much debate on the semantics of word choice and definition, I find the use of the term 'quality' to be more productive than 'excellence'. In particular, it is useful to deploy this term because it can easily be expanded to its plural form which emphasises the role of context and the diverse set of *qualities* that may be important. This is a significant step forward because we can then ask, what are the qualities of important, valuable or well-conducted research that differing localities may wish to adopt and reward.

One of the qualities that is often valued is that of being 'international'. As we have noted, this is conflated with 'prestigious' and 'excellent' when in fact what it most often means in practice is 'North Atlantic'. So-called 'international' journals are not representative, neither in the distribution of authors, nor readers, nor of subject matter.

This observation of the rhetorical conflation of 'international' for 'North Atlantic' offers one way forward.[8] That which is truly of general value for humanity in Western knowledge traditions (i.e. that which is 'Western' but not 'imperial') should be of global value or interest. Where 'international' can be comfortably replaced with the word 'global', this is a signal that something may be of general value. Where this replacement is uncomfortable or inappropriate, it is perhaps a signal that the issues at stake are parochial to the North Atlantic and therefore of peripheral concern for the global community.

Simply banishing the word 'international' from our language – or at least that discourse held in English – may be a valuable way forward. But beyond this, we need to consider how to institutionalise locality in our language. Or rather, local communities need to consider how best to achieve this. Considering how references to local, national and regional interests and needs are valued in contrast to the 'international', and how this is valorised through the choice of terminology and rhetoric, is key.

Social knowledge creation and measuring use and engagement

In other work, I and others have worked with social models of knowledge creation (Neylon 2017; Potts et al. 2017; Hartley et al. 2019). Central to all these models is that knowledge – in the sense of generalisable applicable insight – is made at the boundaries between groups. The Sabato-Botana Triangle model in the context of networks provides a means of defining at a high level what kinds of groups might be of interest, particularly if we expand the three corners to four by including civil society, media and community organisations.

Diversity is a first-order principle in these models and the challenge of knowledge production is supporting institutional and cultural forms where that diversity results in productive interactions. The scaling of knowledge production requires us to seek not just diversity in itself, but an increasing diversity of groups to continue contesting and generalising knowledge.

In the traditional North Atlantic centres of scholarship there are increasingly important sources of diversity in interactions beyond the academy. They come through agendas such as 'wider engagement' and 'citizen science'. In thinking about the qualities that research evaluation and resource allocation should support, a key question is how those choices foster knowledge flows *between* the academy and these communities. Transitional countries, especially those with surviving indigenous and traditional knowledge cultures, have much richer resources to draw on.

The key word here is 'between' and not 'from'. Guided by the Sabato-Botana Triangle, we are concerned with the strength of connections. Enduring and valued connections depend on real benefits flowing to both ends of the line. What is not proposed is a new cycle of expropriation where the only change is that the colonial state is local, but rather that the aspiration is the production of new institutions and cultural forms in which indigenous knowledge holders, local communities and local researchers all benefit from the strengthening of connections. In concrete terms this means expanding beyond traditional citations to ensure that those knowledge flows within the local context and from peripheral to peripheral spaces are tracked, measured and rewarded. Practically, this requires an identification of important communities and a consideration of how knowledge flows between them can be tracked.

One small example of this is the recent description by Peter Dahler-Larsen (2018) of the tracking of citations flowing from non-English literature. This illustrates the use – even the subversion – of the neo-colonial infrastructure to examine different flows. It also illustrates how the process of seeking to track those flows that are not privileged by the neo-colonial infrastructure can be a challenge. Systems to do this effectively will need to be produced or at least configured to address local needs. External infrastructures may be useful, but they need to be assessed and judgements made about the extent to which the systematic biases they create can be addressed and managed.

There are many ways in which citation measures could be tweaked to address the concerns of transitional countries, but they remain citation counts, which render invisible a significant proportion, if not the majority, of global scholarship. New infrastructures will be necessary to support the rewarding of local and periphery-to-periphery information flows. Tracking community engagement offers a useful set of proxies for doing this and signals that these relationships are valued.

The qualities of traditional Western scholarship that are worth adopting and celebrating may be recognisable as those that productively support equitable internal and peripheral knowledge flows. They will be the ones that support effective translation and dissemination of knowledge across the group boundaries that matter. A candidate list might include reproducibility, transparency and effective targeting of communication to the most appropriate audiences. A candidate list to reject might include citation counts, journal rankings and impact factors, the set of problems that they privilege and the frameworks that reinforce those privileged problems.

It is well established that those things which are measured tend to come to matter. While this is almost always framed as a negative consequence, it can also be a powerful means of signaling, provided it is applied thoughtfully and intentionally. By identifying and seeking to evaluate concerns of local importance, and the connections that might successfully address them, these subjects and areas will naturally be privileged in the minds of scholars and the societal discussions in which they are embedded.

The key here lies in identifying and negotiating the set of groups that matter. This does not mean that a total abandonment of the 'traditional' measures of excellence is necessary or even appropriate. The traditional centre of the Western academy is *one* of the groups that matter. Continued interaction to maximise local 'extractive capacity' for knowledge produced in these resource intensive centres is of value. But it is just one group among many. The challenge lies in a process of bootstrapping that local capacity alongside local confidence and above all community *and* state trust in the new institutions that are being formed.[9] This is nothing less than culture building and it is not a simple path, but it is the one that most preserves agency and choice.

Future directions: Taking a global lead

One framing found in the current volume deals with how and whether the Global South can choose to learn from ideas on research excellence that come from the North. I believe that a deeper examination suggests that the opposite position has more merit. What can the traditional North Atlantic centres of research learn from peripheral, Southern, post-colonial and transitional countries' perspectives on what research matters?

Although it may be more impressionistic than strongly evidenced, my experience is that scholars in Southern, post-colonial and transitional contexts bring a much richer understanding than scholars from the North of how to connect scholarship to local societal issues. In Europe and North America, it sometimes feels we have forgotten how to value research of local relevance, regarding it as unworthy of publication, let alone funding.

By contrast, the systems, funders, institutions and scholars of Latin America and Africa have led the world on public access to formal publications, on the building of sharing infrastructures, and in the support of research units that have a deep insight into the societal issues around them (see e.g. chapters in this volume by Barrere, and by Allen and Marincola). While the UK and the Netherlands have loudly promulgated policies and spent vast sums of money on delivering open access, Brazil has had higher levels of open access for a decade and many Latin American universities retain higher levels of open access publishing than comparators in the North. South Africa has higher levels of open access to publications on issues that are the main contributors to South African mortality than the Netherlands.[10]

Latin American infrastructures for data management and sharing are a decade more mature than shared infrastructure in Europe and North America. Southern African infrastructure such as DataFirst leads the world on providing multi-tiered data management and protection. Research organisations, for example the South African Labour Development Research Unit and their programme of 'Impact Dialogues' provide a model for how expertise, informed by transparent evidence, can be debated and engaged with by political and government players in a productive manner.

There is much work to be done. The confidence to support and build on these existing institutions is sometimes tenuous. Brazil has lost its global lead on open access; the vast funding underpinning the European Open Science Cloud may overtake the Latin American capacities of RedCLARA and Redalyc. And, admittedly, frequently these areas of success are found in the richest amongst the lower-income countries, for example, Brazil and South Africa.

Often, these are technical infrastructures, not supported by strong governance institutions and culture. Funding may be highly politicised, fragmentary and unpredictable. The systems, and the connections between those industrial, governmental and knowledge production systems identified by Sabato and Botana, need to be strengthened together. Building the information and technical infrastructures that will allow the observation and evaluation of these connections, while signaling that these are valuable, is a delicate and difficult process. Building new institutions and cultures that privilege local connections will be challenging. More than that, it is an ongoing process; one that is unlikely to ever be finished, but will require ongoing renewal.

But underpinning all this, from my perspective at least, is that the existing institutions and culture of scholars in post-colonial and transitional countries already have a deeper rooted connection between capacity building and local needs. A deeper connection between researchers and the issues of their societies. Even amongst the researchers in the countries tackling problems in North Atlantic ways, with North Atlantic goals of publication in North Atlantic venues, the choice of problem is still guided by an awareness of context. For many researchers in Europe and North America, it feels that the very idea that they should be thinking about local issues is anathema. They must focus on 'excellent' research of 'international' interest.

The old centre has arguably lost its way. In my view, there is an opportunity for those who have been seen to be on the periphery to take the lead, if they choose to do so.

Notes

1 For two examples of quite different arguments along this general line see Neylon (2015) and Frodeman (2017).

2 Although unfamiliar, I adopt the term 'North Atlantic' to avoid the use of 'the North' (which is geographically incorrect, e.g. excluding disadvantaged regions of Eastern and Southern Europe), or 'developed countries' (because it privileges one specific history of 'development'), or 'colonial powers' (because this is often taken to not include the US or Canada). It is a deliberate attempt to localise a specific set of epistemic and evaluative cultures, rather than grant them any sense of being 'universal'.

3 The work to consult in this area is that of Fyfe and co-workers (e.g. Fyfe and Moxham 2016; Fyfe et al. 2017; Moxham and Fyfe 2018), Baldwin (2015a, 2015b, 2017), Czisar (2018) and others. It is a growing area but sparsely populated as yet.

4 Grosfoguel's (2000) critique of dependency theory and its associated political movement, and what I refer to here as a 'lack of confidence' provides an interesting counter. This is similar to the discussion of 'feudalmania' as a more thoroughly worked-out description. However, Grosfuegel would critique the implicit stance of 'developmentalism' in my argument.

5 My focus here is on the post-colonial countries of Latin America and Africa. A large part of the commercial pivot has been towards China as a major new market. While aspects of my argument are relevant to China and other East and South-East Asian nations, the context there is different in important ways that are beyond the scope of this paper.

6 Clearly this is not restricted to corporations focused on research services, but also applies to the global corporate-states of Google, Amazon, Apple, Facebook and also Tencent and Ali Baba, offering a different view on the identity of colonial powers.

7 See for example Ferretti et al's (2018) comment: 'Despite different positions about the controversial underpinnings of research excellence, widely discussed by the majority of interviewees from each of the three categories, none offered slight or indirect suggestions on how to go beyond the issue of quantification of research quality for policy purposes ... [signalling] an inevitable commitment to quantification: when asked about research excellence, different actors tend to digress around specific implementations and their

implications but do not question in a strong manner the overall scope of the indicator as a means to map or ascertain scientific quality.'

8 A challenge with the use of the terms 'North' and 'Northern' and equally here with the adoption of 'North Atlantic' is the place of countries such as Australia and New Zealand with their differing histories of colonisation and independence to those of Africa and Latin America. While beyond the scope of this article, and similar to the previous comment on China and other East and South-Eastern Asian countries, much of the same argument holds. Australian conceptions of research excellence are particularly driven by quantities of citations through the Excellence in Research Australia process and arguably the lack of independent research strategy in Australia stems from many of the same issues discussed here. However, the relative level of historical and structural disadvantage is clearly distinctly different.

9 There is much in common here with Chataway and co-workers' concept of 'embedded excellence' (Chataway et al. 2017; see also this volume). The challenges also parallel the difficulties in addressing the concept of 'Southern Theory' (Rosa 2014). A parallel, but more critical, perspective on the choices in front of 'the South' is found in Thanapal's response to Lorde (Thanapal 2017).

10 Author's own work. The terms 'HIV', 'stroke', 'diabetes', 'heart disease', 'lower respiratory tract infections', 'diarrhoea' and 'road injury' were used as search terms in Web of Science for articles from 2013 to 2018. Articles with an affiliation to South Africa or the Netherlands were selected and the number that was reported by Web of Science as open access was used to calculate the percentage. For each search term, South Africa had a higher percentage of open access reported.

References

Baldwin M (2015a) Credibility, peer review, and nature, 1945–1990. *Notes and Records: The Royal Society Journal of the History of Science* 69(3): 337–352. https://doi.org/10.1098/rsnr.2015.0029

Baldwin M (2015b) *Making Nature: The History of a Scientific Journal*. Chicago: University of Chicago Press

Baldwin M (2017) In referees we trust? *Physics Today* 70(2): 44–49. https://doi.org/10.1063/PT.3.3463

Bornmann L and Daniel H-D (2008) What do citation counts measure? A review of studies on citing behaviour. *Journal of Documentation* 64(1): 45–80

Carayannis EG and Campbell DFJ (2009) 'Mode 3' and 'Quadruple Helix': Toward a 21st century fractal innovation ecosystem. *International Journal of Technology Management* 46(3–4): 201–234. https://doi.org/10.1504/IJTM.2009.023374

Chan L, Posada A, Albornoz D, Hillyer R and Okune A (2018) Whose infrastructure? Towards inclusive and collaborative knowledge infrastructures in open science. *ELectronic PUBlishing.* Connecting the Knowledge Commons: From Projects to Sustainable Infrastructure (June). https://elpub.episciences.org/4619/pdf

Chataway J, Ochieng C, Byrne R, Daniels C, Dobson C, Hanlin R, Hopkins M and Tigabu A (2017) Case studies of the political economy of science granting councils in sub-Saharan Africa. https://idl-bnc-idrc.dspacedirect.org/handle/10625/56808

Chubb J and Richard W (2017) Artifice or integrity in the marketization of research impact? Investigating the moral economy of (pathways to) impact statements within research funding proposals in the UK and Australia. *Studies in Higher Education* 42(12): 2360–2372. https://doi.org/10.1080/03075079.2016.1144182

Chubb J and Reed MS (2018) The politics of research impact: Academic perceptions of the implications for research funding, motivation and quality. *British Politics* March, 1–17: online. https://doi.org/10.1057/s41293-018-0077-9

Connell R (2016) Decolonising the curriculum. *Raewyn Connell (blog)*. October 2016. http://www.raewynconnell.net/2016/10/decolonising-curriculum.html

Csiszar A (2018) *The Scientific Journal: Authorship and the Politics of Knowledge in the Nineteenth Century*. Chicago: University of Chicago Press

Dahler-Larsen P (2018) Making citations of publications in languages other than English visible: On the feasibility of a PLOTE-Index. *Research Evaluation* 27(3): 212–221. https://doi.org/10.1093/reseval/rvy010

Donovan C (2007) Introduction: Future pathways for science policy and research assessment: Metrics vs peer review, quality vs impact. *Science and Public Policy* 34(8): 538–542. https://doi.org/10.3152/030234207X256529

Donovan C (2008) The Australian Research Quality Framework: A live experiment in capturing the social, economic, environmental, and cultural returns of publicly funded research. *New Directions for Evaluation* 2008(118): 47–60. https://doi.org/10.1002/ev.260

Etzkowitz H and Leydesdorff L (1995) The Triple Helix – University-Industry-Government Relations: A Laboratory for Knowledge Based Economic Development. *SSRN Scholarly Paper ID 2480085*. Rochester, NY: Social Science Research Network. https://papers.ssrn.com/abstract=2480085

Ferretti F, Guimarães Pereira Â, Vértesy D and Hardeman S (2018) Research excellence indicators: Time to reimagine the 'making of'? *Science and Public Policy* 45(5): 731–741. https://doi.org/10.1093/scipol/scy007

Frodeman R (2017) The impact agenda and the search for a good life. *Palgrave Communications* 3 (February): 17003. https://doi.org/10.1057/palcomms.2017.3

Fyfe A, Coate K, Curry S, Lawson S, Moxham N and Mørk Røstvik C (2017) *Untangling Academic Publishing: A History of the Relationship between Commercial Interests, Academic Prestige and the Circulation of Research*. Zenodo. https://doi.org/10.5281/zenodo.546100

Grosfoguel R (2000) Developmentalism, modernity, and dependency theory in Latin America. *Nepantla: Views from South* 1(2): 347–374

Fyfe A and Moxham N (2016) Making public ahead of print: Meetings and publications at the Royal Society, 1752–1892. *Notes and Records: The Royal Society Journal of the History of Science* 70(4): 361–379. https://doi.org/10.1098/rsnr.2016.0030

Hartley J, Potts J, Montgomery L, Rennie E and Neylon C (2019) Do we need to move from communication technology to user community? A new economic model of the journal as a club. *Learned Publishing* 32(1): 27–35. https://doi.org/10.1002/leap.1228

Joseph Mbembe A (2016) Decolonizing the university: New directions. *Arts and Humanities in Higher Education* 15(1): 29–45. https://doi.org/10.1177/1474022215618513

Lorde A (1984) The master's tools will never dismantle the master's house. In *Sister Outsider: Essays and Speeches*. Berkeley, CA: Crossing Press. pp. 110–114

Moore S, Neylon C, Eve MP, O'Donnell DP and Pattinson D (2017) 'Excellence R Us': University research and the fetishisation of excellence. *Palgrave Communications* 3 (January): 16105

Moxham N and Fyfe A (2018) The Royal Society and the prehistory of peer review, 1665–1965. *The Historical Journal* 61(4): 863–889. https://doi.org/10.1017/S0018246X17000334

Neylon C (2015) The road less travelled. In: ML Maciel, AH Abdo and S Albagli (eds)*Open Science, Open Issues*. IBICT. http://livroaberto.ibict.br/handle/1/1061

Neylon C (2017) Openness in scholarship: A return to core values? In *Expanding Perspectives on Open Science: Communities, Cultures and Diversity in Concepts and Practices*. Limassol. pp. 6–17. https://doi.org/10.3233/978-1-61499-769-6-6

Potts J, Hartley J, Montgomery L, Neylon C and Rennie E (2017) A journal is a club: A new economic model for scholarly publishing. *Prometheus* 35(1): 75–92. https://doi.org/10.1080/08109028.2017.1386949

Rosa MC (2014) Theories of the South: Limits and perspectives of an emergent movement in social sciences. *Current Sociology* 62(6): 851–867. https://doi.org/10.1177/0011392114522171

Sabato J and Botana N (1968) La Ciencia Y La Tecnología En El Desarrollo Futuro De América Latina. *Revista de La Integración* 3: 15–36

Smith S, Ward V and House A (2011) 'Impact' in the proposals for the UK's Research Excellence Framework: Shifting the boundaries of academic autonomy. *Research Policy* 40(10): 1369–1379. https://doi.org/10.1016/j.respol.2011.05.026

Soudien C (2015) Looking backwards: How to be a South African university. *Educational Research for Social Change* 4(2): 8–21

Sugimoto CR, Robinson-Garcia N, Murray DS, Yegros-Yegros A, Costas R and Larivière V (2017) Scientists have most impact when they're free to move. *Nature News* 550(7674): 29. https://doi.org/10.1038/550029a

Thanapal S (2017, November 16) On Audre Lorde and the Master's Tools. *Djed* (blog). https://djedpress.com/2017/11/16/audre-lorde-masters-tools/

Wouters P (2016) Semiotics and citations. In: R Sugimoto (ed.) *Theories of Informetrics and Scholarly Communication*. Berlin: Walter de Gruyter

PART 2

Research excellence in practice

CHAPTER

7

Utility over excellence: Doing research in Indonesia

Fajri Siregar

Background

Improving the quality of science has been one of the agendas of Indonesia's current regime under President Joko Widodo (Jokowi). Under his guiding development plan, *Nawa Cita,* research and development (R&D) is seen as playing an important part in two points, namely to improve productivity and competitiveness (pillar no. 6 of the plan) and to achieve economic resilience (pillar no. 7).

On several occasions, the president has shared his perceptions on the functions of research in the context of Indonesian development. In his view, research ought to *'rediscover its utility. It should be useful and serve the needs of society. It should strengthen innovation and competitiveness. It should not be done for the sake of research itself.'* [1]

To align the functions of research for the purposes of economic development, Jokowi made the crucial decision to merge the Directorate General of Higher Education (then under the Ministry of Education and Culture) with the Ministry of Research and Technology at the beginning of his reign in 2014. Since January 2015, all science and research-related activities have been officially placed under the Ministry of Research, Technology and Higher Education (Ristekdikti).

Since then, the government has made no secret about its desire to enhance research for development purposes. It has been clearly stated in all official documents (e.g. RIRN, Ristekdikti strategic plan and other official documents of the ministry) that the goal of Jokowi's administration is to increase productivity and competitiveness. Science, especially R&D, would have to act along these corridors.

However, Jokowi's administration is no pioneer in utilising R&D to maximise domestic growth in a technocratic outlook (Amir 2007). If anything, his plans and intentions have only been made more explicit than the previous regimes. In fact, he continues walking on a path that has already been laid out since the early millennium.

National science and technology policy in contemporary Indonesia: A brief overview

Efforts to improve Indonesia's national science policy already began shortly after the political reform in 1998. At the turn of the new millennium, the Government of Indonesia (GOI) laid out plans to decentralise higher education and revamp the national science policy.

The first steps in creating a more coherent science and technology (S&T) framework were laid out in Law No.18/2002, known as the Law on National System of Research, Development and the Application of Science and Technology (OECD 2013). At the time of writing of this chapter, this legislation was being revised to accommodate the most current needs of state-driven innovation,[2] but it essentially covers all matters pertaining to research excellence and the utilisation of science for economic growth. Research downstreaming and valorisation, commonly termed as *Hilirisasi,* is a core idea behind the legislation. The law posits that the central government (then still the Ministry of Research and Technology – RISTEK) plays a coordinating role and has the highest authority in delegating all other roles and functions of the many different actors within Indonesia's science ecosystem.

Institutionally, the National Research Council (DRN) was established in 1984 to identify and define S&T development paths and priorities. DRN was also expected to advise on national S&T policies formulated by RISTEK (OECD 2013). However, its role was supposed to be revitalised upon the introduction of Law 18/2002, as the government sought to streamline R&D activities by setting up national research agendas that should serve as roadmaps for public research institutions and universities to follow suit.

Another breakthrough that took place in this last decade was the introduction of a National Innovation Committee (KIN). KIN was established in 2010 to oversee and coordinate developments across the national innovation system (OECD 2013). However, as an ad hoc institution, the council did not manage to achieve its targets, as it was disbanded towards the end of the then President Susilo Bambang Yudhoyono's reign in 2014. The function of KIN (and the Directorate General of Higher Education) was merged into the Ministry of Research, Technology and Higher Education (OECD 2013).

Persisting institutional challenges remain one of the hurdles in creating an enabling ecosystem for quality research and innovation to flourish, despite notable efforts in streamlining the institutional arrangement. The GOI has taken other measures to improve the conditions of doing science and creating technology, with the aim of enhancing research utilisation at the heart of its plans. These measures include greater freedom to research actors in the planning and execution of public funds for research and commercialisation activities (Brodjonegoro and Moeliodihardjo 2014).

The government's decision to decentralise decision-making in research to R&D actors has created a new tension between autonomy and control. This is exemplified by government policies geared towards more productivity, without creating the necessary preconditions or environment where quality research can thrive (ACDP 2013; Brodjonegoro and Moeliodihardjo 2014). These contradictions will be elaborated in the following sections.

New Public Management and the functions of research

The apparent push of GOI to increase the output and productivity of science and research can be viewed as reforms influenced by a *New Public Management* approach, focusing on increasing efficiency in public organisations (Christensen 2011; Hidayat 2012). This is visible in the structural reforms taking place within public universities, with seven state universities gaining a new legal status through government regulation No. 61/1999. Through the regulation, state universities were restructured into State Owned Higher Education Autonomous Legal Entities (BHMN), which allowed more autonomy in attaining external funding to support their activities (Rakhmani and Siregar 2016). The eventual output expected by the GOI is an increase in quality research and other academic products such as patents, joint collaboration and wider international cooperation.

According to Christensen, New Public Management reforms taking place within the university reflect the more general reform trends in the political-administrative system that are geared towards neoliberal principles (Christensen 2011). In the Indonesian case, increased efficiency that correlates with better output is not only expected from universities, but also from other research institutions using public funds. Local research councils (DRD), the Agency for Technological Analysis and Implementation (BPPT) and the Indonesian Institute of Sciences (LIPI) are other science actors whom the central government expect to increase their outputs (Oey-Gardiner 2011), most notably in the form of international publications.

An increased emphasis on productivity and output is a key feature of the current Indonesian science landscape. If New Public Management is a key notion to understand institutional reforms driven by structural pressure (Christensen 2011) (which parallels with the notion of good governance in other sectors), then globalisation in the form of increasing international standards is the other dominant force.

The role and presence of international agencies in Indonesia plays a critical role in this regard. The World Bank, for example, asserted its agendas through several projects to help shape a more effective S&T sector, such as IMHERE (2005–2012) and RISETPRO (2013–2020). The Australian government, on the other spectrum, has taken part in trying to connect the research to policy nexus through its long-term programme 'Knowledge Sector Initiative', involving other major international actors, such as the Overseas Development Institute and the Australian National University. The United States, through USAID, also took part in the effort of improving the management of higher education through its HELM (2011–2016) project.

Beside these programmes, the Ministry is also well aware of, and well related with, other institutions such as Frauenhofer Gesellschaft, the World Intellectual Property Organisation (WIPO) or initiatives of the United Kingdom (UK) through the Newton Fund and Innovate UK, as well as the Ford Foundation and other donors who have all introduced their respective notions of quality. Collaborative programmes have raised the exposure of international benchmarks to Indonesian academia.

All of the above initiatives have contributed in helping Indonesian researchers understand the notion of quality research, albeit without explicitly conveying the term 'research excellence'. Through programme frameworks and performance indicators, notions of quality and international standards were translated to the Indonesian science community in order to achieve the objectives of the said projects.

Performance assessment and measurement

Adopting international standards in a local context

To measure its own performance in science and technology, the GOI has used several sets of globally accepted indicators. As documented in the official long-term national research master plan (RIRN), Ristekdikti refers to indices such as the Global Competitiveness Index (GCI) and Global Innovation Index (GII) to situate Indonesia's relative position in competitiveness and economic performance (Kemenristekdikti 2018).

Notable indicators that are deemed especially important to measure Indonesia's progress include: Gross Expenditure on Research and Development (GERD), multifactor productivity, a headcount of researchers and researchers-to-population ratio. These are some of the main performance indicators used by the GOI to measure the country's progress in S&T. The Global Innovation Index, in contrast, uses indicators such as knowledge creation, innovation linkages, information and communication technologies (ICT), R&D and tertiary education. The

Indonesian government eventually incorporated a set of six indicators into RIRN, as shown in Table 1.

Table 1: National research contribution targets

National targets	2015	2020	2025	2030	2035	2040
Multi factor productivity (%)	16.7	20.0	30.0	40.0	50.0	60.0
GERD/GDP (%)	-	0.84	1.68	2.52	3.36	4.20
Annual state budget for research/GDP (%)	0.15	0.21	0.42	0.63	0.84	1.05
Total number of researchers (headcount)	1 071	1 600	3 200	4 800	6 400	8 000
Potential researchers (%)	-	20	40	60	80	100
Productivity	0.02	0.04	0.07	0.10	0.14	0.18

Source: RIRN document (2016)

GERD is one of the first indicators Ristekdikti uses to understand the general condition of the research environment. Compared to other ASEAN countries, Indonesia is still behind, allocating only 0.2 % of its gross domestic product (GDP) to research, compared to South Korea, ASEAN and BRICS countries, whilst surpassing only the Philippines (0.1%) – see Table 2.

On top of the universal standards referred to by the Ministry, the GOI also looked elsewhere for an international benchmark. For its long-term development agenda, the country set its sights on South Korea, citing the country's relatively comparable situation in the 1970s, in which both countries endured conditions of low growth. South Korea went on to achieve a much higher speed of development as the country accelerated due to a significant amount of technological contribution and science utilisation. This is what Indonesia aims to emulate.

The case of South Korea has convinced Indonesian policy-makers to pursue incremental yet specific improvements, especially in the realms of human resources and the contribution of S&T towards domestic economic growth (Kemenristekdikti 2016). To take an example, the document stipulates the aim to have a ratio of 1:1 in terms of post-graduate to undergraduate student by 2040, citing South Korea's achievement (Kemenristekdikti 2016).

The availability of human resources is indeed one of the main indicators in science and research. This is why Ristekdikti aims to increase the number of researchers and engineers (perekayasa) that are available to undertake both applied and basic research, especially

Table 2: GERD of ASEAN and BRICS countries

Country	**GERD (%GDP)**
South Korea	4.2
Singapore	2.2
China	2.1
Malaysia	1.3
Brazil	1.2
Russia	1.1
India	0.8
Thailand	0.6
Vietnam	0.4
Indonesia	**0.2**
Philippines	0.1

Source: Kemenristekdikti (2018)

those under the auspices of state institutions. According to the data from LIPI and BPPT, by 2016 Indonesia had recorded a total number of 9 556 researchers and 2 295 engineers. The government recorded a steady increase of researchers and engineers as depicted in Table 3.

The GOI is not only targeting an increase in available scientists, but also students with the potential to become scientists. This is why the ministry also monitors the number of postgraduate students enrolled at higher education institutions. Another indicator for this is the number of international students enrolled at Indonesian universities. The ministry utilises these numbers as part of its stick-and-carrot approach towards the quality management of public universities and as an indicator of internationalisation, which is an important element of competitiveness. Along these lines, increasing the amount of collaboration, both national and international, is another sub-theme of productivity.

The overall goal should, however, not be understood as just becoming on par with South Korea. Above all, the GOI aims to achieve economic competitiveness to become a global powerhouse, citing a McKinsey report that suggests Indonesia's potential to become the seventh largest economy in the world, were it to achieve its full potential (Mckinsey Global Institute 2012).

In doing so, the government bought into the principles of the Triple Helix, believing that in creating a productive science ecosystem, the

Table 3: Growth of engineers and researchers, 2010–2016

Year	Researcher	Engineer	Technical researchers	Nuclear experts
2010	7 502	1 967	N/A	N/A
2011	7 658	2 176	N/A	N/A
2012	8 075	2 176	N/A	419
2013	8 713	2 261	N/A	457
2014	9 128	2 341	2 735	457
2015	9 308	2 332	2 705	437
2016	9 556	2 295	2 499	N/A

Source: Kemenristekdikti (2018)

first step is to align business, academia and the state. As a result, inventions already present in other sectors, such as civil society, were often overlooked (Amir and Nugroho 2013).

In terms of research downstreaming and valorisation, Ristekdikti also tried to be creative. In 2016, Ristekdikti introduced the measure of *Technology Readiness Level* (TKT) as a determinant of funding eligibility.[3] TKT serves as a measurement tool that assesses the readiness of a research project to translate into commercial entities. The introduction of this measure also indicates greater support for research projects with greater commercialisation potential. There is a preference for research that is ready to be made into prototypes, ready to be patented and can be directly applied to commercial purposes.

In this attempt, universities are considered pivotal and the central government is willing to show good faith in its higher education institutions, whilst awaiting a greater return of productivity after more than ten years of structural reforms and financial autonomy.

Translating standards into practice

Macro level

The GOI has introduced policies that push for a coherent framework in improving the nation's science ecosystem. Besides the currently finalised long-term national research master plan (RIRN) that runs until 2040, the government previously referred to National Research Agendas (ARN) developed by the National Research Council.

RIRN outlines the government's research priority sectors and the ensuing budget allocation within the upcoming periods. The document aims to serve as a research roadmap for ten sectors: food, energy, medicine, transportation, information and communication technology, defence, advanced material, maritime, disaster management, as well as social science and humanities. The agendas are set to be coordinated with national development priorities to realign scientific development with long-term economic growth. RIRN itself is being translated into concrete action plans, with the introduction of Ministerial Decree 40/2018 to enforce the programme. The decree also serves as guidance to translate the research priorities into a National Priority Plan 2017–2019.

Meso level

At an institutional level, the ministry has set its sights on operationalising further measurements of S&T development. Hence, further indicators are being developed. This includes an index on regional competitiveness (Indeks Daya Saing Daerah) that charts the capacity of provinces and districts, basically copying indicators used in the GCI and GII indices.

A core component in achieving local competitiveness, which is also an integral component of Jokowi's science development agenda, is the establishment of Science and Techno Parks (STPs). This is a fitting example of how to implement a regime's vision into a workable programme. Due to various factors, the initial target of establishing 100 STPs has hit a bump and is now revised to 66 STPs across the archipelago. Referring to the ten research areas stipulated in the Prioritas Riset Nasional (PRN) 2017–2019, it is clear that food and agriculture is the main theme of STPs to be established.

Another priority programme close to the heart of Ristekdikti officials are the Centers of Excellence[4] (COEs) that are spread throughout several regions across the country. According to Ristekdikti, the goal is to increase institutional capacity, relevance and boost productivity of innovation, especially in the industry sector. The ministry has assisted over 208 institutions spread across universities, ministerial research institutions and industries to cultivate innovative and productive practices as can be seen in Table 4.

Table 4: Number of COEs based on institutional category

Institution	**Number**
Non-ministerial research institution	70
Industrial research institution	13
University research institution	48
Ministerial research institution	77
Total	**208**

Source: Kemenristekdikti (2018)

Predictably, the approach is also on increasing the number of institutions to receive assistance from the ministry. Further evaluation should be undertaken to look at the impact of COEs on increasing local economic growth and whether they contribute to establishing local innovation systems.

Having established macro agendas of research for development purposes, the government went on to tackle issues pertaining to productivity, specifically target-oriented individual improvements. The key issue for the government was how to translate targets into workable programmes or changes in practice.

Individual/micro level

While most research outputs are measured at the institutional level, it is eventually the individual who has to live up to the heightened expectations. It is the individual who has to perform and 'survive' the trappings of the neoliberal academia (Rosser 2016).

Having recognised the low performance of Indonesian scholars internationally, in 2012 the then Directorate General of Higher Education introduced a decree[5] that requires students (both undergraduate and postgraduate) and lecturers to publish in scientific journals. This was followed by a similar decree in 2015, revising the previous rule and focusing on postgraduate students only. This move was not welcomed by the academic community, given their already heavy workload in teaching and also bureaucratic management (Rakhmani 2013).

The government went on to target more senior lecturers who were deemed to be underperforming through the issuance of Research, Technology and Higher Education Ministerial Regulation No. 20/2017. Through this regulation, the government aimed to push middle-level to high-level scholars to publish in journals (specifically, SCOPUS-indexed ones) or else lose their professional allowance.[6] Similarly, lecturers who had already obtained a professorship or Guru Besar (distinguished Professor), were asked to increase their publications output, with the threat of having their professional allowance revoked.

Predictably, the ministry received a public backlash from the academic community, with many scholars writing open letters and op-eds in the media to criticise the move. One notable article written by an Indonesian scholar labelled the mindset as the 'Spectre of SCOPUS' (Mulyana 2017), referring to the government's obsession with increasing the number of publications in international journals, without first improving the quality of infrastructure and providing the necessary preconditions for scholars to be productive.

As part of the public service, lecturers in Indonesia are obliged to comply with the civil servant regulatory framework in order to advance their careers. While some financial incentives have improved over the last years, the many rules and restrictions have hampered their academic freedom and often prove to be a stumbling block in expressing their ideas and aspirations. As civil servants, mobility is restricted and pursuing a postdoctoral position abroad, for example, is officially against the rules once a tenured position at a public university has been obtained (Rakhmani and Siregar 2016; Team 2016).

This is where professional obligations become more apparent and the said 'passion' is put to the test. Junior academics, who have completed their doctorate from an overseas university, and return to an Indonesian university with a relatively respected position, are tasked with juggling between performing academic tasks, while fulfilling managerial duties within the department or faculty, with the latter occupying almost a third of the daily or weekly workload (Rakhmani and Siregar 2016).

Indonesian academics, both junior and senior, are inclined to multitask. Given the relatively low basic income, most scholars are likely to search for additional financial incentives (Suryadarma et al. 2011). By securing a managerial position within the university bureaucracy, an academic adds an important safety net in the form of added take-home pay. Others prefer to occupy themselves with external projects, performing consultancies or policy research that adds financial stability and builds their reputation outside the campus. A majority of social scientists surveyed between 2014 and 2015 were shown to have additional income on top of their regular salary (Rakhmani and Siregar 2016).

Conducting external research is not forbidden, although not actually encouraged. Indonesian academics are asked to adhere to the three principles of academia or *Tri Dharma Perguruan Tinggi*, namely *teaching, research, and community service*. The performance of academics is assessed annually, based on the percentage of those three components. Yet unsurprisingly, teaching is still the dominant component for many academics across regions and universities.

Writing, especially publishing in a scientific journal, seems to be a habit whose virtues are not always understood, particularly by the older generation of Indonesian academics (Rakhmani et al. 2017).[7] To many Indonesian scholars, creating impact is much easier to achieve by writing op-eds and popular articles in the national media. There is a greater sense of fulfilment in being published in a renowned national newspaper (e.g. *Kompas, Jakarta Post)* or in the popular *Prisma* journal than, for example, in the *Journal of Southeast Asian Studies* (Rakhmani et al. 2017). It is therefore no surprise that many Indonesian scholars are not even aware of their own *H-index*, as the thought of publishing in an international journal has never occurred to them (Rakhmani et al. 2017).

The challenge is now to shift the perspective from seeing writing as an obligation to seeing it as an activity that enhances critical thinking and quality improvement within academia itself. That this shift is driven by a technocratic, top-down approach only reminds Indonesian academics how they are still under the control of a government whose commanding attitude is born out of living in the times of a neoliberal spirit.

The term 'publish or perish' applies very much to the Indonesian context, and not only in the Global North, the only difference being the lack of openness in the academic community and meritocracy to embedding research excellence into its organisational culture. This 'quantity over quality' conundrum will likely become the new status quo. It is foreseeable that Indonesia will see an increase in international publications, yet many questions will remain open regarding its real impact on academic quality. Against this background, the pursuit of research excellence will most likely be a by-product of pursuing tangible research objectives, rather than of virtue. Government's push for a more outward-looking attitude is therefore not always a bad policy to have.

Research excellence

Signs of improvement

As of early 2018, Ristekdikti had important progress to proclaim. Based on the latest SCIMAGO data, the ministry highlighted the steep increase in Indonesian international publications in 2017; the number of journal articles almost doubled, especially in the field of natural science.[8] The ministry sees this as an achievement, referring to their persistence in pushing academics to produce more publications, using the stick-and-carrot approach discussed in the previous section.

According to Ristekdikti, the number of international publications has increased, especially between 2016 and 2018. According to official Ristekdikti statistics, the number of international journal articles published rose from 2.057 in 2011 to 8.091 in 2015, having quadrupled within the four years. The average rate of increase was 28.8% for each year. Hence, the general trend of publication is positive. If quantity is seen as a measure of improving the research environment, then Indonesia is doing things right. This positive trend is seen for both national and international journals, as well as conference-based proceedings. The increase in international exposure means that the amount of international cooperation has also increased. This has enabled greater mobility of Indonesian academics (e.g. scholarships and seminar funds). The overview of this rising trend can be seen in Table 5.

In terms of infrastructure, the ministry can also claim to have improved significant aspects of the research environment. An important example here is the setting up of an integrated national publications database of journals called SINTA.

With regard to funding, the ministry has introduced a more flexible, output-based funding mechanism where research is only audited at the end of the research process, according to the pre-agreed output. This will enable easier multi-year funding that has hampered long-term research projects for many years in Indonesia, especially research institutions relying on the annual state budget. Overall, the ministry is still on track in achieving its medium-term goals, as summarised in Table 6.

Seen from a critical point of view, however, these clear-cut indicators set by the GOI do not necessarily illuminate the question of quality. The GOI has also set up national standards on education containing standards of teaching and content of curricula, but these do not reflect nor consider aspects pertaining to research excellence. What are regulated through the national standard are minimum budgetary criteria and allocated commitments for research activities.

Understanding excellence in the Indonesian context

It is fair to say that research productivity and utilisation have been key themes for the Indonesian government. Generally speaking, scholars live in a time where their research is expected to fill a performative function (Lyotard 1984). The issue of the relevance of research to wider societal development has often been highlighted by President Jokowi. Scholars are expected to ask questions of societal relevance and to conform with the common goals of national development, which in all fairness, is not too different from their role during Soeharto's *New Order.*

From a technocratic point of view, the overemphasis on research productivity and utilisation is a necessary step in achieving immediate

Table 5: Overview of journals and conferences, 2011–2015

Year	Journal				Conference			
	International journal articles	National journal accredited by Kemenristekdikti	National journal not accredited by Kemenristekdikti	Total	International conference	Regional conference	National conference	Total
2011	2 057	1 413	10 325	**13 795**	1 843	279	2 438	**4 560**
2012	3 113	1 316	11 521	**15 950**	2 160	430	3 245	**5 835**
2013	4 464	1 232	20 948	**26 644**	4 298	907	5 608	**10 813**
2014	6 459	1 330	19 845	**27 634**	5 037	1 214	6 447	**12 698**
2015	8 091	1 367	18 318	**27 776**	5 989	704	8 116	**14 809**

Source: Kemenristekdikti (2018)

Table 6: Ristekdikti performance indicators

No.	Target indicators	2015	2016	2017	2018	2019
1	Number of registered patents	1 580	1 735	1 910	2 100	2 305
2	Number of published international journal articles	5 008	6 229	7 769	9 689	12 089
3	Number of prototypes	530	632	783	930	1 081
4	Number of industry-ready prototypes	5	15	15	15	15

Source: Directorate General for Strengthening Research and Development (2016)

developmental goals. However, seen from the perspective of a pioneering researcher, the many overwhelming targets result in the feeling of a diminishing space to undertake frontier science or blue skies research. The government is also on the brink of undermining the role of social science by prioritising natural science, both in principles and practice (Rakhmani and Siregar 2016). The government should understand the virtue of doing basic research, or research in the realms of social science and humanities, science that triggers deep dialogues and is about civilisational matters. After all, research excellence is not only about the quality of one's work, but also about whether it brings about a change in paradigms (Kuhn 1996). On many fronts of societal issues, Indonesia badly needs this.

As long as targets come in the form of mere numbers, the achievement of quality will not be the main objective. The academic community is capable of achieving these targets, yet the achievement of excellence will not be inherent in the process. Local standards of quality, utilisation and excellence may also differ from advanced industrial countries and need to be considered. For example, while others are already questioning the effectiveness of peer-review mechanisms, Indonesian academia is still in the phase of firmly embedding peer-review systems into the academic culture. For many actors in academia, it is a process of habituating or of a learning process of building a critical mass to embed peer-review processes.

Conclusion

The Indonesian government has introduced several means of improving its research environment within the last two decades. Macro-level policies as well as institutional changes were introduced to achieve a coherent research ecosystem. The government has made clear that research is an important element of achieving national development targets, with science as an important pillar to contribute to long-term economic growth.

The indicators are clear-cut: economic competitiveness, multifactor productivity, the headcount of researchers, researchers-to-population ratio, gross expenditure on R&D, SCOPUS-indexed articles, citation index, gross contribution towards GDP, and more. The measurement of quality ultimately relies on, and is partly reflected by, these numbers.

What remains to be done is finding a balanced way to achieve these measures. The government has attempted to create a productive research environment by providing the necessary infrastructure. It has also enforced a stick-and-carrot approach, largely informed by neoliberal thinking that has been prevalent in the global sciences and higher education systems. Indonesian universities are no exception to this norm as they gleefully play catch-up in the world university rankings, leaving Indonesian academics little left but to play along.

In the context of an aspiring, lower-middle-income country, research utilisation is more significant than the actual pursuit of excellence itself. For emerging economies such as Indonesia, *Hilirasi* or research valourisation is the main priority. It is seen as a key driver of innovation and is what every major policy and programme revolves around.

Against this background, research excellence should be understood as a by-product of research utilisation. In particular, international collaboration and exposure have helped to raise awareness of matters pertaining to research quality. Understanding the standards of international assessment is an effective measure of shaping 'quality' and 'excellence', which is something the academic community itself should be concerned with, rather than the technocratic, output-driven bureaucracy.

This leaves the Indonesian academia with its own internal homework, namely to build and sustain a local 'critical mass' to habituate the culture of peer review and engender merit-based academia. The challenge for Indonesian academia is to understand the rules of the game, and to incrementally own it. These are matters beyond conventional measurements, but are ingredients of excellence that can elevate the quality of research in Indonesia to enable its scholars to compete on the highest academic playing field.

Notes

1 President Joko Widodo, FRI National conference, 29 January 2016.
2 https://www.antaranews.com/berita/645751/menristekdikti-revisi-uu-sinas-iptek-wadahi-inovasi
3 Research grant recipients can continuously update and monitor the readiness level of their research through an online platform. See https://risbang.ristekdikti.go.id/layanan/tingkat-kesiapterapan-teknologi/
4 See http://pui.ristekdikti.go.id/index.php/beranda_en/profile. Accessed on 18 August 2018.
5 See Directorate General of Higher Education decree No. 152/E/T/2012 and MoRT decree Dikti No 44/2015.
6 See http://www.thejakartapost.com/news/2018/06/10/wanted-6000-new-journals-to-publish-150000-papers.html. Accessed 3 July 2018.
7 See for example the critique written by Franz Magnis Suseno and Dikti di Seberang Harapan. *Kompas*, 8 February 2012. https://edukasi.kompas.com/read/2012/02/09/08343285/Dikti.di.Seberang.Harapan. Accessed 10 November 2018.
8 See https://ristekdikti.go.id/publikasi-ilmiah-internasional-indonesia-terus-melesat-nasir-himbau-untuk-jaga-momentum/. Accessed 3 July 2018.

References

ACDP (2013) *Developing Strategies for University, Industry, and Government Partnership in Indonesia*. Jakarta: ACDP

Amir S (2007) Symbolic power in a technocratic regime: The reign of B.J. Habibie in New Order Indonesia. *Sojourn: Journal of Social Issues in Southeast Asia* 22(1): 83–106

Amir S and Nugroho Y (2013) Beyond the Triple Helix: Framing STS in the developmental context. *Bulletin of Science, Technology & Society* 33(3–4): 115–126

Brodjonegoro S and Moeliodihardjo B (2014) *University-Industry Collaboration*. GOPA

Christensen T (2011) University governance reforms: Potential problems of more autonomy? *Higher Education* 62(4): 503–517

Hidayat R (2012) Politik Pendidikan Tinggi Indonesia Pasca Orde Baru: Reformasi Tata Kelola Dalam Perspektif New Public Management. *Jurnal Komunitas* 6

Kemenristekdikti (2016) *Perbandingan Data dan Indikator Ilmu Pengetahuan dan Teknologi*. Jakarta: Kemenristekdikti

Kemenristekdikti (2016) *Rencana Induk Riset Nasional*. Jakarta: Kemenristekdikti

Kemenristekdikti (2018) *Lanskap Data IPTEK Nasional*. Jakarta: Kemenristekdikti

Kuhn T (1996) *The Structure of Scientific Revolution*. Chicago: The University of Chicago Press

Lyotard J-F (1984) *The Postmodern Condition: A Report on Knowledge*. Manchester: Manchester University Press

Mckinsey Global Institute (2012) *The Archipelago Economy: Unleashing Indonesia's Potential*. Mckinsey Global Institute

Mulyana D (2017, 21 February) Hantu SCOPUS. *Kompas*

OECD (Organisation for Economic Co-operation and Development) (2013) *Innovation in Southeast Asia*. OECD

Oey-Gardiner M (2011) *In Search of an Identity for the DRN*. Jakarta: Knowledge Sector Initiative

Rakhmani I (2013, 25 February) Publikasi Ilmiah Dan Solusi Jangka Pendek. *Kompas*

Rakhmani I and Siregar F (2016) Reforming Research in Indonesia: Policies and Practices. *Working Paper*. New Delhi: Global Development Network

Rakhmani I, Siregar F and Halim M (2017) *Policy Journal Diagnostics Study*. Jakarta: Knowledge Sector Initiative

Rosser A (2016) Neo-liberalism and the politics of higher education policy in Indonesia. *Comparative Education* 52(2): 109–135

Suryadarma D, Pomeroy J and Tanuwidjaja S (2011) *Economic Factors Underpinnning Constraints in Indonesia's Knowledge Sector*. Jakarta: AusAid

Team TR (2016) *Perspectives and Experiences of the Research Culture at Universities in Indonesia*. Jakarta: Palladium

CHAPTER 8

Supporting research in Côte d'Ivoire: Processes for selecting and evaluating projects

Annette Ouattara and Yaya Sangaré

Introduction

Excellence in research is a relevant concept and used in all areas and research structures. However, there is no consensus on the definition of excellence. It has seen a boom in its use in the 2000s and in developed countries that consider it a fundamental criterion for defining their policies in their systems of higher education and research. Even if, unfortunately, there is no consensus on the meaning of excellence, and given the tangled web of fuzzy concepts and ambiguous meanings, there are still attempts to objectify and operationalise the term (Tijssen et al. 2002). However, these attempts have not resulted in the standardised identification of selection criteria, evaluation methods and peer review of projects. Implicitly, excellence can be seen as an effort to achieve the highest possible quality, considering the circumstances. This conceptualisation is rooted in the different research-funding policies, and social and cultural environments in which universities and research centres operate.

The *Programme Appui Stratégique à la Recherche Scientifique* (PASRES) (Strategic Support Program for Scientific Research), is built around the ambition to be one of the best research funding bodies in Africa. For

PASRES, the selection and evaluation of funded projects is first and foremost the pillar that guarantees the reliability and authenticity of any research. This selection is based on criteria and fundamentals that take into account the definition that PASRES gives to the concept of excellence in research. Internal excellence evaluation indicators are identified by the programme in the project selection and evaluation process. Competitive funds for the financing of scientific research and the fruit of the Côte d'Ivoire-Swiss cooperation, it was set up on 15 June 2007. From 2008 to 2018, PASRES has financed a total of 201 projects.

For PASRES, excellence in research must rely on the expertise of researchers and research institutions to produce relevant, rigorous and applicable research results, while responding to different national needs and proposing alternatives to stimulate the development of the country. It must respect the various protocols and ethical processes and rely on collaboration between researchers and institutions. Excellence in research therefore focuses on issues that require an approach based on new, high-quality scientific and technical knowledge. This involves the use of an appropriate and original research methodology. This quest for excellence also requires an allowance for adequate equipment and facilities to research institutions to enable scientists to break the threshold of theory and adopt practices to create a globally competitive scientific system. In addition to this equipment, it would be necessary to provide open access to data and promote North/South and South/South scientific exchanges to our researchers, even if the national language of communication constitutes a barrier to scientific exchanges and to the integration of those researchers in various research networks. In addition, in order to meet international standards, the quality of research produced by Côte d'Ivoire researchers must be improved, even though the scarcity of research grants still constitutes a real obstacle.

But, beyond this purely academic vision, PASRES believes that excellence in research in African countries must transcend the publication stage of research results in scientific magazines and journals. It must include the 'research uptake' dimension. This mainly scientific enrichment (publications in scientific journals and magazines) in African countries must be simultaneously economic, cultural and social. Since science is at the service of development, the results of the research conducted must consider national, regional and international realities, depending on the funding the project receives. As stipulated by Yule (2010), the scope of research quality dimensions must include the utility, accessibility and quality of end-user impacts.

In order to achieve this, PASRES has (based on the Swiss National Science Foundation (SNSF) model) identified and adapted to the national context, a number of criteria and activities likely to initiate this excellence in research in Côte d'Ivoire. We thus have the process of evaluation and selection of projects, as well as the capacity-building activities of PASRES laureates, researchers and instructor-researchers from universities, research centres and Côte d'Ivoire's polytechnic university.

This chapter aims to present the perception of excellence in research at PASRES through its selection and peer-evaluation process of funded projects, and to draw some reflections that may inform a research-granting agency in a low- or middle-income country.

Excellence in research and the selection process for funded projects

At the end of a call for projects, the projects (submitted by a tandem composed of one or more university researchers and one or more community partners) are evaluated by the PASRES Scientific Council (SC). The evaluation process is based on criteria related to the scientific and social relevance of the programme's objectives, the involvement of partners, student training, knowledge mobilisation and feasibility (schedule and budget). These criteria, approved by the PASRES Steering Committee (SC), are contained in a project call for all researchers and partners.

Twice a year, the Executive Secretariat of PASRES opens a call for proposals, which must meet the formal requirements as specified by PASRES. Projects are pre-selected on the basis of eligibility criteria and then evaluated by a jury of experts set up by the programme. Each project is evaluated by a minimum of two experts from the identified research area. The maximum amount that can be granted by PASRES per eligible project is 15 000 000 XAF (approximately 23 000 Euros). Given the limited amount of grant money awarded per project, PASRES strongly encourages project co-financing. With this in mind, the participation of the programme in consortia such as ERAFRICA, Biodiversa-EU and the Belmont Forum help national researchers to obtain greater funding and expand their social capital through the networking system.

Calls for thematic projects are launched by PASRES once a year to finance specific research projects responding to a specific national priority. This call is mainly in line with the programme's strategic objectives and responds to emergency situations that weaken our social and economic environments. These include issues related to climate change and agriculture (especially food).

Funding awarded on excellence criteria

Within the framework of the funding of research proposals, the selection of research projects to be supported benefits from the expertise of specialists from the scientific world, who know the needs and the urgencies of the research in the ten fields financed by PASRES.

The needs of researchers cannot be 'compartmentalised'; PASRES receives requests from all areas of research each year. Thus, in order to proceed with the selection of the best proposals, PASRES has identified criteria based on its vision of the quality of research that transcends disciplinary boundaries.

These four main criteria are taken into consideration in the evaluation of project proposals submitted to PASRES. The first criterion is concerned with administrative aspects, including a letter of commitment from the principal researcher of the project, and a letter of intent from the institution hosting the project. A project summary sheet constitutes the first step in the submitted project selection process. The analysis is carried out by the Executive Secretariat of PASRES.

The second level of analysis considers the scientific aspects of the project. As for the content, it relates to the scientific experience of the project leader in the field concerned, the coherence and logic in the development of the project, the collaborations envisaged within the framework of the project's realisation, the attractiveness of the theme, the quality of the project's summary, the presentation of the problem, the definition of both the general and specific objectives, the proposed methodology and the mechanisms for the enrichment of the research results.

The third criterion considered by PASRES is the project's impact. At this level, the PASRES bodies ensure that the projects selected for funding perfectly meet the Sustainable Development Goals (SDGs), contribute to the fight against poverty, have social, economic and cultural impacts, offer potential for popularisation and transfer, and can be sources of scientific and technological innovation.

The fourth criterion, which assesses the financial aspects, must comply with the following standards:

- Consistency with the nature and duration of the project;
- Respect for the structuring indicated in the submission file;
- Justification of the amounts indicated in the various headings;
- Adequacy of the materials and equipment requested with the submitted project; and
- Cost of realising the realistic and reasonable project not to exceed the maximum amount that may be granted by PASRES. This specific point is one of the eliminatory items when analysing the administrative compliance of the proposal.

Stakeholders involved in the project selection process

The project selection and evaluation process at PASRES involves various stakeholders whose respective roles are as follows:

- *The Executive Secretariat,* which is the centralising and receiving body for the various financing requests and deals mainly with monitoring how the various proposals are handled.
- *The Scientific Council* is responsible for evaluating project proposals by considering the expertise of external evaluators and distributing them according to their level of excellence. Ranking

is based on the scores obtained. In order to limit the subjective nature of the evaluations, the Scientific Council also arbitrates the opinions submitted by external evaluators. They make financing proposals to the Steering Committee to acquire the financing decision.

- *The Steering Committee,* meanwhile, reinforces the funding proposals made by the Scientific Council on the basis of two criteria that consider the national priorities in terms of socio-economic and cultural development and budget availability at PASRES.

Monitoring and evaluation mechanisms, actions, and research excellence at PASRES

Peer review is the main evaluation mechanism for funded projects and articles published by PASRES. The funded projects vary in duration and are regularly evaluated to determine the proper use of the allocated funds and especially the quality of the results obtained.

Monitoring and evaluation mechanisms at PASRES

The monitoring and evaluation mechanisms at PASRES are based on excellence in project management. Indeed, criteria such as: the relevance of the obtained results, adequacy of the methodology, respect of the time chart and the socio-economic and cultural impacts of the project are essential indicators to ensure the continuity of cash outlays. The main evaluators are experts from the scientific world (to judge the quality of the research performed), the private sector (in the context of economic enrichment) and other institutions (to measure the cultural and social potential of the research).

Funded projects may be suspended or permanently terminated in the event of weak or missing results and the misuse of funds. The deviation from the pathway and the objectives set initially within the framework of a financed project cause the project to stop. This weakness is felt mainly in the writing of scientific projects and articles, which is at the base of the lower level of scientific publications in peer-reviewed journals. Côte d'Ivoire does not have enough scientific journals, and to correct this deficit (and in the quest for excellence in scientific publications), PASRES decided to make available two peer-reviewed scientitific journals to researchers and their laureates. PASRES has, to its credit, a social sciences and linguistics journal called *RSS PASRES* (since 2013) and another in environment and biodiversity called *REB PASRES* (2016). Edited periodically, the journal *RSS PASRES* has four issues per year, and the *REB PASRES* has three issues per year. Publications in these journals are entirely at the expense of PASRES (from the peer review to the physical production of the journal). All journal expenses are incumbent on PASRES in order to allow PhD and postdoctoratal students to disseminate obtained research results to the national and international scientific communities. These journals are not exclusively geared towards to national researchers. The programme aims to establish other scientific journals that will take into account the other scientific fields considered by PASRES. PASRES' current challenge is the publication of these journals, from a printed format to an e-journal, as well as their indexation, and greater impact, at the international level. This evolving process responds to the need for a better distribution and popularisation of research results for better visibility of the authors at the scientific and international levels.

In addition, PASRES grants financial and technical support to researchers for the publication of scientific articles in international journals. To this end, training workshops are organised and fully funded regularly by PASRES for teaching scientific writing to researchers in national universities and research centres.

Actions and research excellence at PASRES

In view of its integration into globally accepted standards for research excellence, PASRES has initiated a series of actions within its research management system.

First, PASRES organises monthly training sessions for masters and postdoctoral students at universities and at research centres in writing eligible scientific projects and scientific articles, according to international standards and the recommendations of large donors. These capacity-building activities aim to make the research results of Cote d'Ivoire researchers internationally competitive.

Second, thematic conferences are also organised by PASRES to present the funded research results to the private sector and to civil society. All the actions are fully funded by PASRES. Striving to make excellence central, the programme does not skimp on the means to support and assist researchers, laboratories, universities and research centres. Excellence goes hand in hand with the concepts of quality, efficiency, applicability and competitiveness, and therefore merit awards are granted by PASRES to the best researchers in the scientific fields. These include the PASRES Young Inventor Award, the PASRES Young Parasitology Researcher Award and the PASRES Young Researcher Award in Sociology/Anthropology. Granting of these awards is based on a very selective process, following a call for applications at universities and research centres.

Conclusion

To make research excellence in Côte d'Ivoire a reality, PASRES attempts to instill a culture of excellence in research in national research institutions. Among other things, this involves legislating laws on research at the national level, the development of partnership networks (collaboration), the involvement of the private sector in the execution of research projects, the establishment of an autonomous structure of results appreciation, and the promotion of interdisciplinarity, without forgetting support for mobility, research unit equipment, and gender equality management in terms of research opportunities. This is not easy to achieve because of the complexity of its conceptual definition, but the challenges faced by the Research Granting Councils must motivate the pursuit of excellence in all organisations.

Wishing to help the competitiveness of Côte d'Ivoire researchers, PASRES integrates partnership networks in the field of research and in both regional and international initiatives. This opening of the programme should allow new knowledge and technologies to be acquired at the international level in the fields of science, research, technology and innovation in Côte d'Ivoire. But what about research excellence in our universities and research centres? Exploring this question will help to assess the various perceptions of excellence at the national level for the adoption of a reference framework document.

References

Tijssen R, Visser M and Van Leeuwen T (2002) Benchmarking international scientific excellence: Are highly cited research papers an appropriate frame of reference? *Scientometrics* 54: 381–397

Yule M (2010) *Assessing Research Quality*. International Development Research Centre Peace, Conflict and Development Program. Ottawa: IDRC

CHAPTER 9

Sustaining research excellence and productivity with funding from development partners: The case of Makerere University

Vincent A. Ssembatya

Introduction

Makerere University was established in 1922 as a technical college with an enrolment of 14 students who were all male. In 1949, the university became a University College affiliated with the University College of London, offering courses leading to general degrees of the University of London. This affiliation lasted until 1963 when the university became one of the three constituent colleges of the University of East Africa, alongside the University of Nairobi in Kenya and the University of Dar es Salaam in Tanzania. Makerere University became an independent University in 1970 by an Act of Parliament of the Government of Uganda.

Makerere University had a student population of about 32 000 students as of June 2018, having grown from an average of 3 700 students in the 1970s, 4 700 students in the 1980s and 10 000 students in the 1990s. Figure 1 presents the trends over the last 40 years. The stretch in enrolment had its pinnacle in the 1990s on account of massive education reforms in the country that ushered in universal primary

Figure 1: Student enrolments at Makarere University, 1975–2015

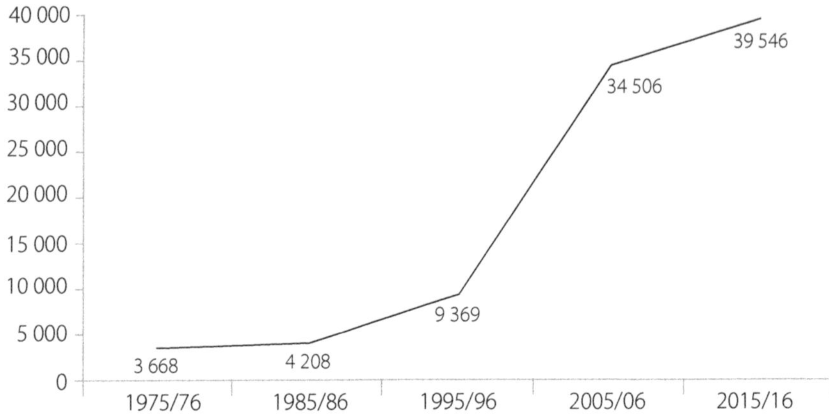

education, leading to a surge in enrolment in primary and secondary schools. The reforms in primary and secondary education made these levels of education more accessible through the introduction of universal primary and secondary education. Alongside these reforms was the liberalisation of higher education (HE) in Uganda, providing for admission of self-sponsored students in government-supported institutions, as well as the introduction of private universities. A new Act of parliament was promulgated in 2001, providing for the creation and regulation of universities and other tertiary institutions in Uganda (Republic of Uganda 2001). This resulted in a growth in the number of universities in Uganda from three universities in 1989 (Makerere University, The Islamic University in Uganda and Mbarara University of Science and Technology) to more than 40 in 2018. Enrolment in universities in Uganda grew from about 10 000 students in 1990 to more than 185 000 students in 2015.

Makerere University has about 17% of the enrolment in all universities in Uganda, and about 53% of the enrolment in public universities in Uganda (NCHE 2006). From a time when the university was the only one in the country for a period of over 60 years, this represents a tremendous interest in HE provision by other actors. Initial interest in providing higher education came from faith-based organisations such as the Catholic church, the Anglican church and Islamic organisations, which established universities run by the respective umbrella bodies.

With the introduction of new universities, Uganda's HE system became more diverse and complex, partly due to the growth in the number of public and private institutions and multiple stakeholders with diverging interests; this occurred in the context of the increasing realisation of HE as a key driver in economic development. The introduction of the National Council for Higher Education in 2001 to regulate the HE sector was an essential move by government to have a coherent mechanism of provision of HE training amidst the complexity that had emerged. The regulatory body put in place statutes for a quality assurance framework, largely specifying quantitative requirements for setting up universities, in addition to accrediting curricula at universities. The regulatory body made only minimal mention of research in universities in the range of criteria for regulating universities and other tertiary institutions.

In the reforms that ensued with regard to the devolution of the provision of HE, Makerere University experienced severe perturbations. First, it was to cede human resources, not necessarily through formal arrangements. Such a formal arrangement would include secondment of top administrative staff to start off universities, a cost-neutral arrangement amongst government-funded institutions. An alternative arrangement would involve mentorship from an existing institution, hosting an office of a new university for a couple of years. No such arrangement would work with non-government-funded universities. In the case of such institutions, more aggressive mechanisms were employed, for example offering higher remuneration packages or otherwise attracting human resources away from Makerere University. In any of these scenarios, Makerere University was to let go of vital human resources, since the new universities often sought productive ones as well.

Second, Makerere University had to succumb to duplication of its curricula through non-formal arrangements by informal cooptation of individual members of staff. Such duplication in curricula would lead to a reduction in the numbers of potential students seeking admission in key disciplines at the university, in turn affecting the research capacity in those disciplines.

Third, the university became a victim of smear campaigns in a *zero-sum game* of recruitment; for a new university to attract students, an existing university must lose students. The current HE situation in Uganda is characterised by very erosive primary and secondary education systems. Of the population of pupils who join the first grade of primary education (about 2 million pupils), about 30% will complete the last grade of primary education (the seventh grade); only 5% of these pupils will make it to the last grade of secondary education (senior six), from whom universities draw their enrolment. The current net enrolment in higher education in Uganda is below 10% (much lower than the sub-Saharan average of 16%) and, strangely, quite comparable to that of the nations within the East African community.

Whatever is holding back HE enrolment is doing so harmoniously across the East African region. The growth of universities in Uganda, therefore, is heavily stifled by the inefficiency of the delivery pipeline that begins at the first grade of the education system, and even earlier in early childhood development, which is largely considered to be for the rich and operates mainly in urban settings in Uganda.

Even though many upcoming universities knew that attracting students was to be a tall order, they also realised that getting qualified academic staff to work in their institutions was even more difficult. The only comfort was that since there were no students, the question of hiring was irrelevant. As such, this tailspin plunged these institutions into some form of inertia. It does not seem, as it stands, that it is a lucrative business to set up a new university in Uganda. The whole university system seems to be massively connected; a radical reform from the norm is very risky due to this connection. Such a connection has raised questions around the purpose of the HE system, the trade-off between private and social returns and, most importantly, whose responsibility it is to fund that system. In various forums, private universities have issued an outcry for government to appropriate funding to defray the high cost of their investments. Many of these universities have clearly stayed away from science, engineering, technology and mathematics (STEM) educational programmes. Most of them have glaringly avoided investments in research.

Repositioning to a research-led university

Makerere University took a stand during the strategic planning period from 2008 to 2018, to reposition itself and focus more on research and graduate training, having realised that most of the new and upcoming universities had no capacity for scientific research. Besides, research seemed only to make sense within the framework of government agencies. Even though Uganda was receiving a lot of foreign direct investments, most of those agencies were deploying ready-made solutions emanating from research done elsewhere. There was neither a need nor a compulsion for local content in research. In addition, at the national level, there was a lack of a National Research Council, or a comparative framework with overarching responsibility and capacity, to propel the generation of research for national development. Makerere University realised that it could leverage its ambient position and human resource capacity to recast its efforts to areas where the other universities had limited access. In addition, it made sense for the university to train a pool of academic staff that would potentially be hired by other universities. Leading up to the planning phase of 2008–2018, there was evidence of a growing interest from development partners to support research at Makerere University.

Makerere's strategic choices for the period 2008–2018 not only led to increased graduate student enrolments (initially), but also to increased research outputs (see Table 1). The number of PhDs graduating in a year increased from 30 in 2009 to 75 in 2017 (see Figure 2). In addition, there was a significant increase in the number of research publications by academic staff, as indexed by the Web of Knowledge database, up from 325 publications in 2008 to 944 in 2017 (see Figure 3); this doubled the rate of publications from 0.32 publications per academic staff per year in 2008 to 0.64 publications per academic staff per year in 2017. The National Council for Higher Education (NCHE), in one of its instruments for regulating universities and tertiary institutions, expects academic staff at universities to publish a minimum of one such publication every two years. Research productivity is considered the distinguishing factor between universities and other tertiary institutions that the NCHE regulates. Universities that do not publish are considered glorified high schools. This requirement is one of the dichotomies associated with research, as there is no explicit funding formula that considers the number of publications from the universities. As such, there is no explicit consequence to an already accredited university if it fails to comply with this requirement. This would work best if a university risked losing part of its funding by failing to produce the requisite number of research publications; equivalently, universities would motivate more research by passing part of the publications-generated funding to individual researchers as incentives.

Table 1: Makerere University performance statistics, 2009–2018

Year	Students	Staff	Publications	Staff:student ratios	Publications by staff	PhDs awarded
2009	34 850	1 362	430	25.6	0.32	30
2010	33 112	1 130	495	29.3	0.44	39
2011	33 470	1 236	461	27.1	0.37	46
2012	37 137	1 236	546	30.0	0.44	42
2013	41 122	1 256	554	32.7	0.44	60
2014	42 508	1 398	639	30.4	0.46	66
2015	38 586	1 405	788	27.5	0.56	62
2016	39 546	1 420	819	27.8	0.58	75
2017	31 802	1 470	944	21.6	0.64	69

Makerere University maintained the top share of academic staff with PhDs in the country for the period that ensued. The number of staff with PhDs increased from 469 in 2008 to 790 in 2016. While complying with the requirements of the national regulating body, the university fared quite well within the local region as far as research was concerned. The attention paid to research, and associated scientific knowledge generation, attracted a lot of funding, especially from the OECD countries. The Ugandan government also realised that the numbers of students from its neighbouring countries were increasing, attracted by the high rankings that Makerere University was receiving and so their numbers within the universities in the country were increasing. This influx of students too was a new source of foreign income. As a result, the government was willing to listen to the aspirations of the university, backed up by evidence of its improved performance.

Research productivity at Makerere University: Key policies

PhD degrees for lecturers

Since the year 2000, Makerere University has had a requirement that every lecturer hold a PhD degree. The only exceptions are the clinical medicine disciplines and those lecturers who were already serving in the university system in 2000. The School of Law had argued for a similar exception to this requirement, but this was denied. For the medicine discipline, it was successfully argued that a PhD was not a requirement for the medical profession and that insisting on this requirement would hurt the university by limiting access to the practising physicians, who would otherwise offer service in the medical school. In the ensuing years, it has paradoxically emerged that the most prolific publishers are the non-PhD staff in medical disciplines. The medical disciplines contribute more than 45% of the research output in Uganda. No similar evidence exists in any other disciplines in the university.

The NCHE has modified the requirement for a PhD in universities to allow for the hiring of registered PhD students who are progressing normally. The modification to the PhD requirement was compelled by the difficulties of attaining a sufficient number of PhDs. It is estimated that Uganda has about 2 000 PhDs, amidst a requirement for over 10 000 PhDs (UNCST 2011). The current PhD deficit is over 8 000 PhDs. This deficit cannot be covered with the current production rate of about 100 PhDs per year (Makerere contributes 75% of the country's production).

This new development, however, is likely to alter the trend at which PhDs have been acquired at the institution. This is one of the dichotomies associated with harmonising requirements for running universities, especially those funded by government. The NCHE is

Figure 2: Number of PhD graduates at Makerere University, 2008–2018

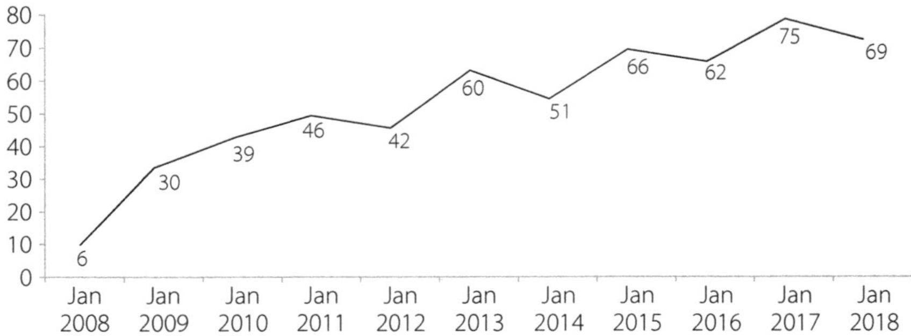

Figure 3: Makerere University research publication output

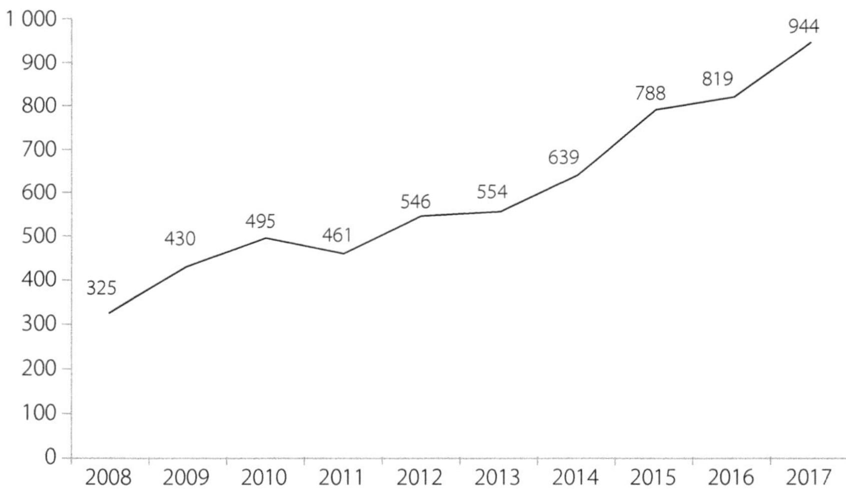

Source: Web of Knowledge database (2008–2017)

often forced to lower standards to accommodate all universities under its purview. Setting high standards tends to cause a retraction in the capacities of the majority of the universities. For the bigger benefit, however, this requirement could have been a necessary gambit in the growth impasse associated with the higher education sector.

Staff appointments and promotion policies

The Makerere University appointments and promotion policy requires academic staff to publish a set number of publications (in peer-reviewed journals) for appointment or promotion to the different categories in the academic staff establishment. The university runs five distinct ranks for its academic staff: assistant lecturer, lecturer, senior lecturer, associate professor and professor. Moving from one rank to another has distinct requirements and attracts several incentives. For instance, to be promoted from lecturer to senior lecturer, one needs three publications, whereas five extra publications are required to move to the subsequent level, in addition to teaching experience and service to the community. The promotion policy is one of the fundamental drivers in motivating the production of publications, as promotion carries monetary incentives.

The appointments and promotion policy requires the supervision of graduate students to completion (a varied mix of masters and PhDs) for senior academic positions in the university establishment. For instance, to be promoted to the level of associate professor, one is required to have supervised at least one PhD student up to completion. This is in addition to other requirements such as publication in peer-reviewed journals. Only senior lecturers may supervise PhD students. The supervision requirement has supported the acceptance of supervisory roles by academic staff for more than one reason. Ironically, it is not directly financially lucrative to supervise students as very little extra pay is associated with the effort that goes into the activity. In addition, the completion rates at PhD level are at about 6%; there are no guarantees and one could easily end up with empty hands. In the case of Uganda, graduate training is currently broadly for fee-paying students. In the 1980s, and before, all university education was free. Students now struggle to cover costs amidst increasing costs of education. The cost of a PhD in Uganda averages around USD 50 000. It is not surprising that students will opt for PhDs that offer the best opportunity for timely completion, as well engagement after the doctoral studies. Quite often development funders avail scholarships in chosen areas of study.

Publication output production from PhD theses

At Makerere University every PhD student is required to publish two research publications in peer-reviewed scientific or scholarly journals (or at least have these papers accepted for publication) before they can graduate.

In all the above-mentioned policies, these publications are a common currency to assess performance – both productivity and quality. The quality dimension is connected to the *peer-review process* applied by these journals to assess submitted manuscripts. As it is often difficult to determine the quality of a research publication, for any set of publications submitted for different administrative considerations, a number of expert committees are required to assess the submitted publications for the associated purpose. Oftentimes the vetting process is considered time-wasting, especially since these publications would have gone (or are considered to have gone) through an elaborate quality assurance mechanism put in place by publishing houses and editors of journals. It goes without saying that some journals may skip this rigourous process, thereby leaving much to be desired in this role. On occasion, the university has sought post-publication opinion on some publications submitted for the purposes of promotion.

A common critique associated with the heavy emphasis on producing these publications is that little emphasis tends to be placed on activities associated with *good teaching*. As a consequence, teaching is (potentially) less incentivised and often simply taken for granted. The implemented 'publication-biased' policies have tended to create a calibre of staff who are 'too good to fail'; those staff that have already produced high numbers of publications may have no immediate needs to show ongoing excellence.

There is no real incentive to improve quality at the top rank of professor; this raises the question of what would drive publication output, and other research quality considerations, at this level.

To address this issue, the university decided to implement the position of *emeritus professor* for those who have reached the mandatory retirement age of 70 and yet continue to exhibit high performance within their disciplines. This position does not attract salaries from the university. An emeritus professor has access to university facilities at the same level as that of an ordinary professor. In addition, the university expects its emeritus professors to attract funding from which they may derive financial benefits. So far, there has been no dash to become an emeritus professor. The likely explanation is that universities that are not government funded are free to employ professors of any age. As a consequence, professors have opted for paid positions in privately funded universities, as opposed to remaining as volunteers in public universities.

The dilemma: Funding from development partners

With the ever-increasing aspiration for excellence, and globalisation pressures amidst shrinking resources available for its operations, Makerere University has partnered with a range of development partners in an effort to diversify its resources. The tuition stream of income is inadequate especially in the face of low gross domestic product (GDP). Most of the excellence measures are not corrected for GDP variations; this is depicted in international rankings. There would have been efficacy variables that correct for innovative utilisation of limited resources to generate reasonably comparable outputs. It is very cost-effective for development partners to spend funds on research in lower-income developing countries where the cost of living is relatively low. The average output per invested US dollar is certainly higher in the high-income countries. This fact is also a cause for a dichotomy when a project is bilateral and results are to be reported in both economies.

In the period 2000–2012, Makerere University received more than USD 214 million from development partners, mainly for research (see Table 2). The current annual donor operational budget (about USD 3 million a year) is about 6% of the university's total budget. Most of this budget is dedicated to research. Component research funding comes to the university through the Government of Uganda's 'Initiative for Science Support' that operates under the office of the President of the Republic of Uganda. About USD 2 million per annum is allocated to specialised projects identified by the president under the Presidential Science Initiative. Without support from the development partners, research at the university would barely be possible. The long-term research arrangements have supported the building of institutional capacity to do research by supporting PhD training. These collaborations with foreign partners and funders have also helped networking researchers who would otherwise be isolated. It is estimated that more than 50% of the PhDs obtained between the year 2000 and the year 2010 were acquired from outside Uganda, with support from development partners. The Swedish government alone has supported the training of about 300 PhDs in the period 2000–2015 with the utilisation of the 'sandwich mode' of training, where the student has supervisors from all countries partnering in the project.

Table 2: Development partners' research funding to Makerere University, 2000–2012

Funding agency	United States dollars
Government of Sweden	62 380 000
Government of Norway (NORAD)	39 809 385
USAID	28 926 924
Rockefeller Foundation/IDA/WB	24 468 824
Carnegie Corporation of New York	16 591 000
European Union (EU)	9 992 885
CDC	5 670 572
African Capacity Building Foundation	5 150 000
Netherlands Government (NUFFIC)	4 750 000
IDRC	4 073 651
DFID	3 621 209
Ford Foundation	2 826 000
Millennium Science Initiative	2 134 453
World Health Organization	1 288 325
Uganda National Council for Science and Technology	1 245 898
Johns Hopkins University	766 228
MacArthur Foundation	735 000
PHEA (Partnership for Higher Education in Africa)	450 000

A dilemma associated with obtaining funding from development partners is with the alignment of the research focus, which tends to be biased toward interests supported by the funder. For instance, in the period 2008–2016, about 40% of the research indexed in major databases was in Medicine, with an additional 8% in Immunology and Microbiology (see Table 3). Whereas Health Sciences and attendant problems are key to Uganda's economy, Agriculture is the mainstay of the economy, employing 40% of the labour force and generating 25% of the country's GDP. Research in Agriculture and Biological Sciences only accounted for about 12% of the total volume of research during the period. Competitive calls for research funding, which emanate from funding agencies in the Western countries, are typically thematic with themes aligned to the intentions of the funder. Such calls also require that partners on the research teams are drawn from countries from the West. These stringent requirements tend to outpace any other considerations that may bring excellence to the table. The rule of thumb is for there to be sufficient overlap in the aspirations of the partnering institutions in the funding collaborations. To exacerbate the problem, capacity-building research has tended to bias the capacity towards the same areas of Western priority, which now pushes the problem of misalignment to the distant future. Some of the capacity-building efforts have lasted for a period of 30 years, leading to the graduation of PhDs in those areas, and the creation of research labs and facilities. In the case of Makerere, the Rakai Health Sciences Program in Southern Uganda, the Iganga-Mayuge Demographic Surveillance Site in Eastern Uganda and the Institute for Infectious Diseases at Makerere sprang out of capacity-building research in the recent past. In essence, the ripple effect that these research centers will create will drift the research generated over time in the same direction for years to come, after the funding agreements have been extinguished.

According to data available in the Scopus database, about 40% of Makerere research output for the period 2008–2016 (3 441 publications) was in the general subject area 'Medicine', with an extra 8% (702 publications) in the area 'Immunology and Microbiology'. Indeed, most of the funding in research is concentrated in Makerere's College of Health Sciences. Some development partners have recognised the possible harmful effects of this bias and have therefore relaxed funding requirements that are now also targeted towards institutional capacity building, as well as supporting the research agenda of the university. The

College of Health Sciences has the mandate to engage research areas and topics that seem to attract the interest of the international community, especially in the OECD countries. The designation of the Sustainable Development Goal (SDG) 3 on health and well-being has bolstered this arrangement. A proliferation of interests is heavily embedded in this SDG and is likely to outweigh all SDGs in terms of investments.

Table 3: Makerere University research publication output by subject areas, 2008–2016

	Subject area	Research publications	Percentage
1	Medicine	3 441	39.5%
2	Agricultural and Biological Sciences	1 039	11.9%
3	Immunology and Microbiology	702	8.1%
4	Social Sciences	686	7.9%
5	Biochemistry, Genetics and Molecular Biology	624	7.2%
6	Environmental Science	405	4.7%
7	Computer Science	206	2.4%
8	Pharmacology, Toxicology and Pharmaceutics	168	1.9%
9	Engineering	155	1.8%
10	Psychology	147	1.7%
11	Veterinary	138	1.6%
12	Nursing	128	1.5%
13	Business, Management and Accounting	109	1.3%
14	Earth and Planetary Sciences	109	1.3%
15	Mathematics	98	1.1%
16	Arts and Humanities	95	1.1%
17	Economics, Econometrics and Finance	88	1.0%
18	Energy	61	0.7%
19	Chemistry	50	0.6%
20	Multidisciplinary	45	0.5%
21	Physics and Astronomy	39	0.4%
22	Neuroscience	36	0.4%
23	Health Professions	34	0.4%
24	Materials Science	34	0.4%
25	Chemical Engineering	26	0.3%
26	Dentistry	22	0.3%
27	Decision Sciences	16	0.2%

Source: (Scopus database, 2008–2016)

Regardless of the interests of the Ugandan government, (as articulated in the country's vision and its development plans), as long as research funding from development partners continues at the current level, Makerere's research portfolio will be tilted towards the interests of those funders. This is likely to remain the case even when the calls for funding seem to imply responsiveness to Uganda's national agenda.

Going forward: Resolving dichotomies from the global economy

Until the mid-1990s the role of HE in Africa's socio-economic development was fairly anomalous; the majority of the education development projects focused either on primary or secondary education. International donors and development partners regarded universities, for the most part, as institutional enclaves neglecting the particular development needs of African communities. However, current research shows that the returns on investment from higher education are not only on the increase, but also surpass those of the other levels of education. There is evidence that countries that have expanded HE systems, with higher levels of investment in research and development (R&D) activities, have higher potential to grow faster in the globalised knowledge economy. It is also evident that research productivity from African universities is under the radar, with Africa's visible contribution at only about 2% of the global research volume; this is at severe variance with the population proportion of 17%. The experience from Makerere University points to the fact that the national economies are yet to mobilise their flagship universities to actively support national development agendas through knowledge generation. As a result, such universities resort to sources of funding for research; these sources may not necessarily take kind interest in those development agendas. Excellence and quality in such cases will have dichotomous readings, one from the funder's point of view and the other from the recipient's. A clear way around this dichotomy is for governments to appropriate funding for key areas to their development agendas.

Uganda's participation rate in HE, as measured by the HE gross enrolment ratio (GER), is about 10% lower than the world average of 26%. According to the 'State of Education in Africa' report (Africa-America Institute 2015), returns to investments in higher education in Africa are 21%, the highest in the world. However, the enrolment rates at universities in sub-Saharan Africa are among the lowest in the world. In the same report, it is also noted that African countries have allocated, on average, 18.4% of government expenditure to education, with Uganda's current allocation at 11% (financial year 2016/2017) down from 16.2% (financial year 2009/2010). The proportion of this budget allocated to HE is about 12% (rather than the recommended 20%).

Another issue relates to the funding allocation model of universities, which is largely dependent on student numbers. Research outputs are not included in this. As such, universities can avoid doing research. To mitigate this disincentive, there is a need to create a National Research Council that appropriates research funding to universities. The same council could devise a research-rating mechanism for professors, as well document and incentivise research efforts in universities.

Development partners play an important role in correcting these historical imbalances which have relegated universities in Africa and the Global South into positions of dismal contributions to the global research footprint. Besides, there is adequate motivation for partners from high-income countries to associate with counterparts in the Global South for further collaboration and synergy in resolving global challenges of hunger, absolute poverty, energy, climate change and health. In addition, the Global South harbours crucial resource reservoirs that are of much interest to researchers from anywhere in the world in the quest for solutions in health and agriculture, as well as in the provision of raw materials for industries. It is crucial that support from development partners be made less stringent for sustainability, and less in terms of offloading international burdens onto unsuspecting populaces and more in partnering for solutions of mutual benefit. Support to development agendas that have been articulated by the regional consortia, countries and the recipient institutions through their research agendas could be a good start.

Finally, the capacity of universities to participate successfully in high-quality research and scientific knowledge generation needs to be increased. Whereas quality (fitness for purpose) research would have to face the question of articulation (in the national and university visions and agendas), an even bigger question would have to be faced in the form of research capacity, research process and resourcing. Currently there are less than 50 researchers per one million people in Uganda, compared with more than 7 000 researchers per one million people in Sweden and over 8 000 per one million people in Israel. Raising this low base requires addressing several institutional, logistical and infrastructural obstacles at various levels throughout the Ugandan educational system. These hurdles range from school inputs, teachers, curricula, long distances to schools, feeding, parental support and examination policies. Other high-priority issues are in the incentives to investments in schooling or returns to investment in education. The government, as the leading provider of social services, has a vital role to play in leveraging HE capacities and outcomes in order to generate the knowledge and skills that are required for economic development and prosperity. The development partners can only play a complementary role in this process. With regard to improving the quality of research, a beginning would be to allocate a reasonable percentage of the GDP (say 1%) to research, improve research organisation and production capacity, strengthen research infrastructure and facilities, regularly review and update the national research agenda, and monitor its implementation through compelling mechanisms to ensure that targets are met.

References

Africa-America Institute (2015) *State of Education in Africa Report, 2015: A Report Card on the Progress, Opportunities and Challenges Confronting the African Education Sector*. Africa-America Institute

National Council on Higher Education (NCHE) (2006)

Republic of Uganda (2001) *Universities and Other Tertiary Institutions Act, 2001*. Entebbe, Uganda: Government of Uganda

UNCST (Uganda National Council of Science and Technology) (2011) *The Careers and Productivity of Doctorate Holders (CDH) Survey in Uganda Report (1990–2010)*. Kampala, Uganda: UNCST Science and Technology Policy Coordination Division

CHAPTER 10

Southern conceptions of research excellence

Suneeta Singh and Falak Raza

Introduction

Research-based inquiry has, and will remain, a process that is integral to how we go about our lives; it kicks into play when we 'research' a holiday or the options for dental treatment. Yet these instances of 'desk research' do not rest upon research rigour.

Quality becomes critical when 'scientific research' is done with the intention of proving discoveries, determining paradigms or changing ways of doing that have the potential to affect the lives of many. Because such research done with these intentions has far-reaching consequences, it is necessary to safeguard its attention to scientific quality and probity. Research funders have, understandably, a particular interest in 'research quality'. They can see the need for research quality and are held accountable for it. With funds for research coming under stress, the need to operationalise the notion of 'research excellence' is increasingly acute.

The difficulty is posed by differing views on what excellence of research means. Tijssen and Kraemer-Mbula (2018) note that 'the underlying generic concept of "research quality" is not so easily pinned down: it is [a] complex, multidimensional notion with many context-specific and time-dependent attributes'. A growing body of scholarship suggests that discussions of research excellence are dominated by the 'Global North', and calls for the knowledge gap between the Global North and the Global South to be addressed.[1] While these terms no longer accurately represent the geographies that they did when coined, they continue to express a social cognisance of the gap between countries that have large-scale influence and those whose influence is more local.

Background

In 2012/2013 Amaltas[2] carried out an enquiry into how Southern researchers view research excellence and how their experiences could inform framework(s) for assessment of research excellence at IDRC (Singh et al. 2013). The aim of the study was to analyse and summarise the prevailing discourse on questions such as: where is the field moving; what are and who are the different proponents of key debates; and what is the spectrum of definitions and approaches being used?

The nature of the study on which this paper primarily rests was exploratory. Its respondents, drawn from the databases of IDRC and Global Development Network grantees, were well experienced and were engaged in multidisciplinary research. The study received responses to a survey questionnaire from over 300 Southern researchers based across the Global South, and in-depth interviews were held with ten researchers identified as being 'innovative' by agencies funding research for development. Over three-quarters of the respondents to the main survey of the study had been born and resided in the Global South, but a majority had completed their last degree in the North, blurring the line between what is a 'Southern' view and what is not.

Since research funded by research councils such as IDRC typically occupies the space of use-inspired research, or research for development, this paper looks at research excellence in the context of use-inspired research. Use-inspired research occupies the Pasteur's Quadrant, a model and term coined by Donald Stokes in 1997 (Stokes 1997). He placed 'quest for fundamental understanding' along one

Figure 1: The Stokes Typology of Research (1997)

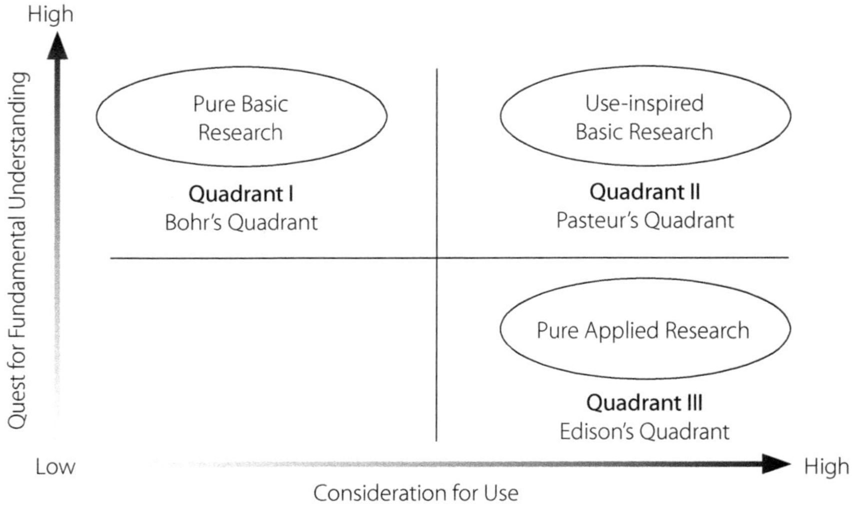

Source: Adapted from Singh et al. (2013)

axis and 'consideration for use' along the other axis in this model (see Figure 1). This paper discusses research that falls into Quadrant II, represented by Louis Pasteur, whose work was the epitome of high quest for both fundamental understanding and consideration for use.

In addition to the work cited above, this paper also draws from experiences across the range of work at Amaltas, in particular a project commissioned by what was then Research Councils UK to identify leading institutions actively engaged in research in the areas of public health and well-being (Amaltas 2015). Over 4 500 peer-reviewed papers were captured by the study. The paper relied on bibliometric analysis using publication and citation counts to capture researchers and institutions working in the identified themes. Institutions were ranked based on aggregation of data of the researchers affiliated with them, using a natural inflection point in the data to classify their institutions as 'leading' or 'other'.

Research excellence and Southern perspectives

Introduction

A significant body of research that takes place in the South is quite often funded by the North; it is natural that the concerns that dominate quality judgements of North-funded research are applied to this research as well. And even if it is not, Southern research is often held to Northern standards and notions of quality when time comes to publish. Yet Northern and Southern researchers operate in markedly different social, economic, cultural and political environments.

Do notions of research quality and research excellence travel well across these different milieus? Do the noisy debates about the definition of research excellence and appropriate indicators for it resonate in Southern corridors? Do those grappling with real-world research in the South find the standards, methods and dimensions that are applied to Northern research persuasive when applied to research in the South? Is the world – both North and South – agreed upon what constitutes research excellence? And finally, is there a level playing field between the North and the South?

Based on the 2013 study and others since, it seems that Southern views on research excellence can broadly be categorised into three brackets: (i) Southern value systems; (ii) dissonance in measurement applied to use-inspired, real-world research; and (iii) the disadvantage that research in 'other' institutions/'other' languages faces. The next subsections address each issue.

Importance of Southern value systems

Our sample of Southern researchers exhibited a wide range of value systems when they define or describe research excellence. Definitions of excellence least frequently identify the traditional dimensions of research rigour, namely, research publications and citations (see Figure 2). Stakeholder involvement, originality and dissemination appear more frequently than publication and citation counts. But exceeding these, by a factor of 2–3 times, are less traditional

Figure 2: How respondents defined 'research excellence'

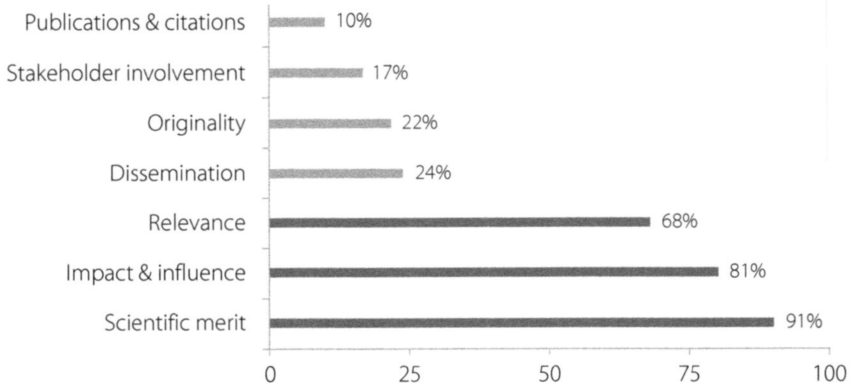

Source: Singh et al. (2013)

dimensions: relevance to clear development needs within the context in which research is undertaken and its impact and influence among key stakeholders. At the top of the list is scientific merit, signaling the central desire to see that research meets the standards of quality and probity that researchers of the South hold.

Southern researchers lay a great deal of stress on the notion of relevance. It is important for them that research be relevant to the country context. They believe that relevance is made more meaningful by ensuring that the research question(s) are framed by communities whose lives were sought to be changed. Thus, *relevance 'to whom'* arises as an important issue to be examined by a research excellence framework. One researcher said, 'Excellence as a uni-dimensional quality is useless for evaluating research. What we need is criteria that incorporate a variety of dimensions of how research can be useful.'

Southern researchers also emphasise that all the possible kinds of influence and impact that the research might have on practice or policy must be taken into account. For Southern researchers, impact is significantly linked to 'other-than-academic' effects as also 'other-than-policy-changing' effects. A researcher noted, 'More robust mechanisms for peer review should be developed; impact on the field of research must be prioritised; public impact should be considered widely rather than being restricted to policy influence.' Thus effects such as gender

Figure 3: Aspects of research excellence that are emphasised by evaluations

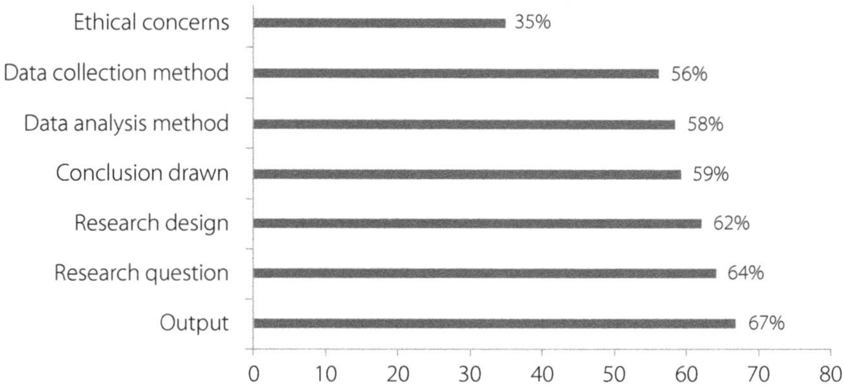

Source: Singh et al. (2013)

sensitisation of field workers or the incorporation of new indicators into a monitoring information system are seen as important by Southern researchers.

Funders of research have highlighted the significance of relevance as a key parameter of research excellence. When Southern researchers are asked to identify what the most important funders of their work emphasise, their responses indicate that funders, *with the exception of Research Councils*, emphasise relevance the most.[3] This runs counter to intuition – it would seem that science granting councils would be most likely to support relevance as an important parameter of well-conceived research.

Southern researchers weigh up scientific merit, influence and impact, and relevance in what they feel research excellence embodies. Yet the practice of research evaluation does not appear to lay as much importance on these dimensions (see Figure 3). Research quality frameworks in use most commonly cover aspects such as research question, research design, methods for data collection and analysis, ethical considerations, outputs and conclusions drawn (Singh et al. 2013).

Researchers conducting 'implementation research' caution against using a common set of dimensions, without taking into account the nature of their research. As one researcher observed, 'I would like evaluators to use different criteria in the evaluation of academic and policy writing…'. They emphasise that research with practice or policy at its heart dealing with complex ideas ought to be judged differently from research for academic outputs.

Given the often-mercurial nature of the environment in which Southern researchers function, adopting a rigid framework with the hope of achieving standardisation is problematic. Use-inspired research conducted in dynamic settings is very likely to deviate from the original research plan. Such complexities and evolving situations in the real world that impinge on research of issues are inadequately captured and/or addressed by available evaluation frameworks of research excellence.

Dissonance in measurement applied to use-inspired, real-world research

Interestingly, despite their emphasis on relevance and influence and impact, Southern researchers are not able to articulate how the dimensions they consider important could be measured. When asked to identify what indicators ought to be used, they fall back on indicators such as publication and citation counts.[4] Is the disjunction in indicators and dimensions of research excellence due to the high value that is attached to research publications in the academic world? Or is it perhaps related to the difficulty of constructing objective and easy-to-apply indicators that can be used to assess impacts?

The reliance on these 'bibliometric' measures to assess research excellence is problematic, given the widely held views on their limitations. Donovan (2007) suggests that although these counts may be a good measure of productivity or impact on subsequent academic publications, these measures do not capture the quality of the papers or the research that lies behind. Citations could be made to advance or refute findings of the cited paper, citation counts may be inflated when research is published in an established journal or under-represented when published in non-English language journals (Jarvey et al. 2012).

From the perspective of use-inspired Southern research, these limitations have implications for how the quality of research is judged.

Due to the context in which Southern research often takes place, researchers may adopt innovative methodologies, often emergent in nature, which have never been used in the North. These are not always valued when assessing a paper for inclusion in a top-notch international scientific journal. The evaluation and publication of multi/ inter/ transdisciplinary work and emerging disciplines – another hallmark of research in the Global South – poses another set of difficulties. Top-tier disciplinary journals are sceptical of publishing interdisciplinary research, and there are few journals that publish interdisciplinary research exclusively (Kulkarni 2015). An OECD report (1997) similarly points to the neglect of grey literature – often of cardinal importance in interdisciplinary work and for innovative developments – in favour of codified literature in scholarly journals that has been a drawback in research evaluation. This is a key concern for Southern researchers, especially because the overwhelming majority of them in Pasteur's quadrant are engaged in multi/ inter/ transdisciplinary work (Singh et al. 2013).

Disadvantage of research in 'other' languages and/or 'other' institutions

Southern researchers are doubly disadvantaged with respect to the language they use for reporting vis-à-vis their native languages. Not only do they have to overcome the hurdle of communicating in a language that frequently does not instinctively 'come to them', the value that they create in their native language is often not assessed as a 'product' of their research. Knowledge products in the local language are ordinarily not taken into account when judging quality of research – this is especially unfortunate as these products might exert considerable influence on local practice and policy, which is the purpose of use-inspired research.

Singh et al. (2013) note that 58% of the respondents in their study have native languages other than English, Spanish and French, the dominant languages of the world. Yet, approximately 85% of the respondents use English, Spanish or French to report results within their own countries and 99% use these languages to report outside of

Figure 4: Most Southern researchers opted to report in English outside their countries

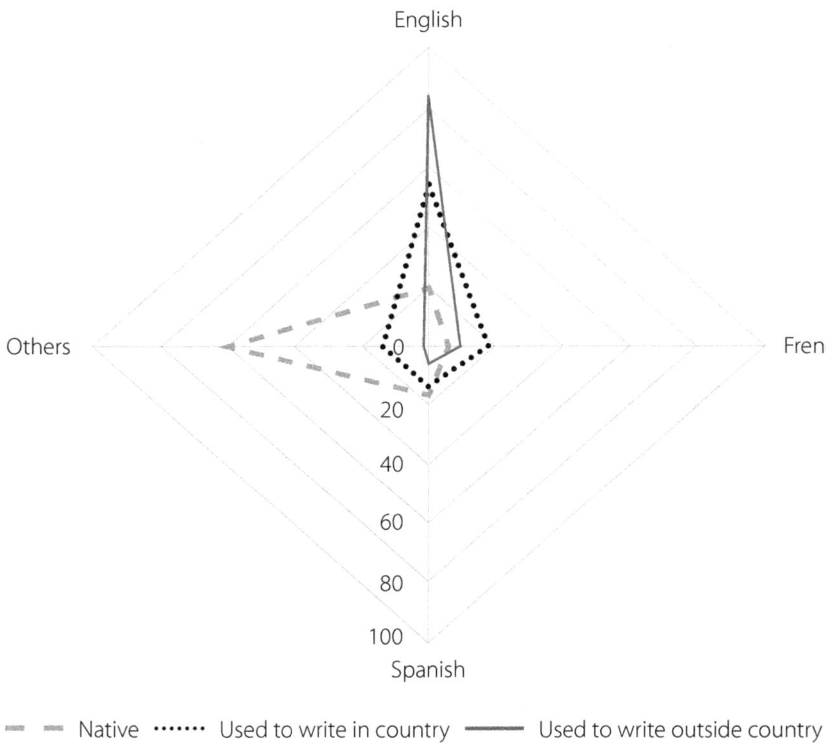

Source: Singh et al. (2013)

their countries. About 83% of reporting outside the country was in English alone (see Figure 4), compared to the 19% who have English as their mother tongue. This raises important questions about the bearing that use of dominant languages has on researchers' facility to report, and the possibility that acceptance of their reports may be prejudiced by their language skills in a language not their 'own'.

The disadvantage to researchers speaking and writing in a language other than English is apparent in their under-representation in counts of publications or citations. Donovan (2007) notes that standard citation counts such as Thomson Scientific have a relatively low representation of regional journals, small research fields and non-English papers. In a vicious circle, academic platforms for native language reporting struggle to survive in a competitive world as they are scarcely cited, discouraging authors from submitting high-quality research articles to these journals (Fung 2008).

The bias *towards established researchers* as evidenced by their bibliometric counts has also been identified by Amaltas (2015) in its mapping of researchers, institutions and funders engaged in the areas of public health and well-being in India. The study found that researchers in 'leading institutions' have higher productivity (publication counts) and greater scholarly influence (citation impact) than researchers in 'other institutions'. Leading institutions are clearly able to nurture researchers to not only publish more, but to also publish in more impactful publications. The obverse – that researchers from other institutions are at a significant disadvantage in terms of publications and citations – also holds true. Some observers have referred to the importance of this kind of professional 'social capital' by relying on collaborations with the North to enhance their reputation.

Future directions

Research excellence encompasses a wide range of meanings. At one level, research excellence is a methodological construct of ensuring that scientific rigour is maintained and that processes which are required to be undertaken receive due attention. At another, and one might argue a more important level, it is a 'political construct' delineating the complex relationship between that being researched and the wider world.

Use-inspired research must be judged on the basis of this construct. Any discussion of its quality must account for the relevance of the research to local issues, the applicability of research findings to the context, and the influence and impact that is generated on the lived reality of the people whose lives it seeks to transform.

There is as equal a need to focus on research process issues, as on the dynamics between the protagonists in use-inspired research and their wider context. Research quality is epistemological, Southern researchers argue, while research excellence is concerned with results

Figure 5: Research excellence encompasses research quality

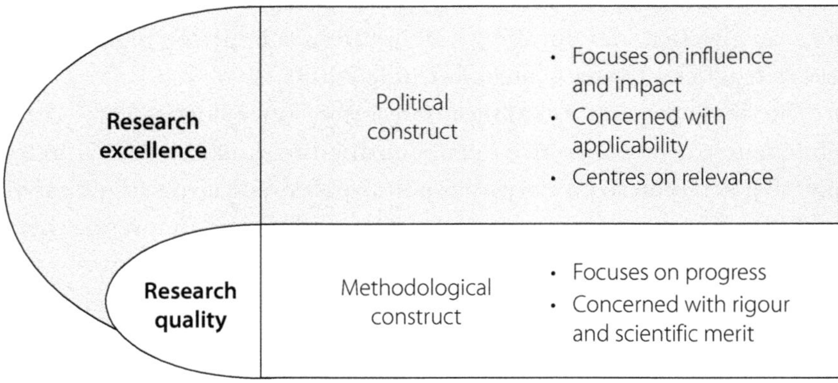

Source: Adapted from Singh et al. (2013)

and application to a policy discourse (see Figure 5). They suggest that it is important to be 'inclusive' and involve those whose lives are sought to be changed in defining the research question in an initial step, and hence the idea of relevance 'to whom' is another important dimension of excellence.

Emergent research methods, and cross-disciplinary approaches are an important facet of use-inspired research; this makes research publications in top journals less likely. Oddly, much of Southern research is undertaken without explicit ethical review, despite it being use-inspired; this may at least in part be related to the many scientific disciplines that are jointly engaged on it. In particular, cross- inter-transdisciplinary work faces pressure from each discipline to conform to its own normative expectation; it also faces an internal problem during the research, as researchers from different disciplines and conceptual frameworks attempt to work together cohesively as one unit.

While policy-making or practice influence is an important aspect that Southern researchers emphasise, there are apprehensions about the delay between taking action and seeing change. Concepts such as 'knowledge creep', what constitutes innovative research, what the most fitting endpoint of the research should be, and the degree of control that researchers have to ensure a desirable impact, go into this mix of concerns. Another important concern is related to accounting for non-English (or Spanish or French) reporting from non-native speakers. Finally, it is also important that the dynamic nature of the context in the case of use-inspired research is factored in.

Conclusion

It is evident that the perspectives of the Global South on excellence in the context of use-inspired research are distinctive from, and somewhat dissonant with, prevailing majoritarian views. Southern perspectives are disadvantaged in so far as their value is judged and consequently, their reach and influence is compromised. Because of this, it is important that science granting councils based in the Southern world ensure that they take cognisance of Southern understanding of what excellent research is; they develop systems that can assess quality in this light; and they elevate the value of Southern-based scientific research.

In considering Southern understanding and systems for assessing quality, an initial step would be to think about the unit of assessment. This has implications for the design of the assessment framework. Various approaches have been taken – department-level assessments are promoted by the Higher Education Funding Council of England (Hubble 2015); fields of research by the Australian Research Council (2018); while the IDRC's Research Quality Plus (RQ+) framework aggregates quality from the projects, upwards to the portfolio level (Ofir et al. 2016). Granting councils will need to make practical decisions regarding the methodology applied, based on availability of evaluators, technology reach, etc.

It may also be worthwhile to consider different performance measures and indicators for the phase of the research cycle when assessing a piece of research work or a research proposal. Given the stress that Southern researchers lay on relevance, outcomes and impact, this becomes particularly critical, bringing focus to the two 'ends' of the research cycle not usually covered in normative quality assessments. Singh et al. (2013) suggest three phases at which quality evaluation could be carried out: (a) pre-grant phase which assesses the conceptualisation of the grant; (b) grant phase which is aimed at understanding the scientific merit and research rigour; and (c) post-grant phase which can gauge influence and impact.

Amplification of the 'voice' of Southern researchers must be an important aim of the work of South-based granting councils. Southern voices must be heard more at international conferences. They must exemplify the kind of research that Southern researchers value. Excellent research and researchers conducting excellent research must be identified and encouragement provided to bring more use-inspired learning to the fore. Institutions must be encouraged to develop high-quality portfolios of work. Finally, granting councils must work closely with leading scholarly journals and academic associations to discuss and put forward examples of what, by Southern lights, is excellent research.

Notes

1 SGCI/IDRC Workshop Concept Note: 'Perspectives on research excellence in the Global South'.
2 Amaltas Consulting Pvt Ltd is a Southern development institution that works to expand intellectual capital and innovative approaches for development.
3 Other dimensions emphasised by funders, as perceived by researchers, were rigour of design, methods of analysis, outputs, dissemination, policy impact, originality, stakeholder involvement and ethics. Academic impact was ranked the lowest.
4 Some of the other indicators suggested by the respondents were changes at policy and community level, relevance of topic, use of innovative design or methods and capacities built.

References

Amaltas Research Council UK, India (2015) *Public Health and Well-being: Mapping Institutions, Researchers and Funders in India*. Amaltas

Australian Research Council (2018) *Excellence in Research for Australia: 2018 update*. Government of Australia

Donovan C (2007) The qualitative future of research evaluation. *Science and Public Policy* 34(8): 585–597

Fung IC (2008) Citation of non-English peer review publications – some Chinese examples. *Emerging Themes in Epidemiology* 5(12). Retrieved on 18 June 2018. https://www.ncbi.nlm.nih.gov/pmc/articles/PMC2570362/pdf/1742-7622-5-12.pdf

Hubble S (2015) *2014 Research Excellence Framework*. Retrieved on 19 June 2018. http://dera.ioe.ac.uk/22708/1/SN07112.pdf

Jarvey P, Usher A and McElroy L (2012) *Making Research Count: Analysing Canadian Academic Publishing Cultures*. Toronto: Higher Education Strategy Associates

Kulkarni S (2015, 4 November) Interdisciplinary research: Challenges, perceptions, and the way forward. Retrieved on 8 June 2018. https://www.editage.com/insights/interdisciplinary-research-challenges-perceptions-and-the-way-forward

OECD (Organisation for Economic Co-operation and Development) (1997) *The Evaluation of Scientific Research: Selected Experiences*. OECD

Ofir Z, Schwandt T, Duggan C and McLean R (2016) *Research Quality Plus: A Holistic Approach to Evaluating Research*. Canada: IDRC

Singh S, Dubey P, Rastogi A and Vail D (2013) Excellence in the context of use-inspired research: Perspectives of the global South. Canada: IDRC

Stokes DE (1997) *Pasteur's Quadrant: Basic Science and Technological Innovation*. Washington, DC: Brookings Institution Press

Tijssen R and Kraemer-Mbula E (2018) Research excellence in Africa: Policies, perceptions, and performance. *Science and Public Policy* 45(3): 392–403

CHAPTER 11

From perception to objectivity: How think tanks' search for credibility may lead to a more rigorous assessment of research quality

Enrique Mendizabal

Introduction

Academic research is undergoing a transformation across the world. Few are the research communities where the pressure to achieve and, most importantly, to demonstrate non-academic impact, is not on the rise (Villanueva and Mendizabal 2016). In some cases, this pressure is regulated and part of national science and education policies. In other cases, where domestic funding for research is wanting, this pressure is enforced by changes in the international development sector which have focused greater attention on measures of value for money and impact.

However, this pressure to have and demonstrate impact has not been matched by changes in the academic sector or, more practically, in the way that universities generate and communicate evidence. By and large, researchers in universities are still judged, for better or for worse, by publication in top academic journals which have only a relatively small effect on non-academic impact. This raises several dilemmas that universities face globally, including in relation to their intended roles in society.

In this chapter, it is put forward that think tanks, which are more often than not judged by a subjective assessment of their credibility, rather than the objective assessment of the quality of their research, offer a rich portfolio of alternatives for universities interested in maximising the non-academic impact of their research. Whether by adopting some of their practices or working in partnership with them, universities may have their cake and eat it, too. In other words, influence a layperson and inspire the experts. Research excellence need not be compromised.

This is partly because, as I argue, think tanks are beginning to successfully establish closer, deeper and more sustainable relationships with multiple (and often new) audiences in a quest to gain credibility. In these new relationships, it is possible to pay greater attention to more objective indicators of quality.

However, to achieve this, it is first necessary to understand research excellence from the perspective of think tanks and, then, consider how different forms of communication for impact may be employed.

It is not my intention to argue that universities ought to be more like think tanks. Nor is it to advocate for a narrow understanding of research that focuses on providing solutions to the questions posed by others – policy-makers, businesses, etc. I am also not suggesting that all universities are equal and I recognise rich regional and national heritages that explain the diversity of the sector across the world. This is diversity, if also found among think tanks. Yet, in almost every context, think tanks and universities overlap and define themselves in relation to each other.

The interaction between think tanks and universities makes it possible for the latter to learn from the former and adopt certain practices that may help them address, in particular, the challenge of assessing the quality of the research.

In the next section, I explore the concept of think tanks, research excellence and credibility to situate it in relation to academic research centres. In the following section, I consider some research and communication strategies that think tanks are pursuing which promote the development of trust with their different audiences. Finally, I put forward a series of implications for research centres and researchers.

Background

Think whats?

The literature on think tanks is heavily influenced by the notion of *waves* or *traditions* put forward most prominently by Diane Stone (Stone and Denham, 2004; Belletini 2007). The former refers to three waves of think tank development: from a few state-centric centres (often set up by governments themselves), to more diverse think tank communities with greater links outside the government and national borders, to, finally, a situation where think tanks are, in essence, acting transnationally.

The concept of traditions refers to regional or national characteristics or development moments that may be helpful in the study of think tanks. Various authors have adopted these ideas: such as Orazio Bellettini, James McGann, and even I, for instance, with Ajoy Datta and Nicola Jones (Mendizabal et al. 2010). However, these notions do not fit nicely with what we find in reality: it is possible to find countless examples of stark differences between think tanks in the same regions and countries –as well as similarities between them across diverse contexts. In a review of think tanks in Latin America, I found several co-existing traditions depending on the origin of the organisations (Mendizabal 2012): be it from non-governmental organisations, academia, the government or other public bodies, and aid-funded projects or networks, for instance.

Moreover, the idea of development *waves*, particularly the suggestion that think tanks are now acting transnationally – more so today than they ever did – more closely reflects the reality of think tanks in developed nations than in developing ones. It also contradicts the evidence: Chinese think tanks, from their very beginning, have been oriented to learning about the world outside China (Mendizabal 2016); Chilean think tanks would not have been possible without support from foreign funders, universities and think tanks (Puryear 1994); and the metaphors that have inspired and driven the formation of think tanks in developed countries have played central roles in think tanks' national histories across the developing world. In other words, this transnationality is by no means new.

An alternative to the study of think tanks is to combine these ideas with perspectives of how politics work – and the roles that different political actors, including think tanks, play – within each polity.

This approach yields some interesting results when we look at cases across the world:

- Elitist, statist and pluralist political systems can play key roles even within the same country (and region) over time;
- Individual organisations can also be driven by more than one of these forces throughout their history;
- Waves of formation or development then cannot be expected to follow a particular linear trend – i.e. increased openness or *transnationality* – but reflect much more complex internal and external forces at play in the spaces that think tanks inhabit;
- Political and economic liberalisation, often assumed to be drivers of think tank formation and responsible for the so-called second wave, are in fact not necessary conditions for the emergence of think tanks;
- Even during periods of autocratic and military rule, think tanks can find fertile ground to develop – and they may in fact be drivers of change; and
- There are several important similarities between think tanks in extremely diverse contexts, which calls into question the relevance of studying think tanks within geographic regions, or even in the imaginary 'developing world' or the 'Global South'.

Out of this emerges an increasingly rich picture in which no single model of a think tank or a single approach to characterise their research, communication and capacity-building efforts – not even in a single country – easily applies.

As a crude oversimplification (Mendizabal 2013), some think tanks have emerged out of academic environments and follow strict academic rules; ever so eager to see themselves as universities without students.

But academic think tanks are not all the same, either. Some maintain strong connections to universities, often hosted by them; others are membership-based organisations which researchers use as contracting vehicles.

Think tanks have also emerged from advocacy or activist communities and therefore pay greater attention to the communication of existing or new research, partly via the development of narratives and discourses. These are rather common in Eastern Europe and the Western Balkans, where think tanks emerged from human rights groups and NGO activism around the fall of the Berlin Wall. There are many think tanks based within government that act as boundary workers between the fields of research and politics.

To complicate the picture further, there are also, increasingly, new private sector think tanks founded by consulting firms, large corporations and business groups. They can carry out high-quality research and cutting-edge communication, even if questions about their intellectual autonomy remain.

As a consequence, or as a possible explanation for this diversity, there is no law that regulates what a think tank is – or cannot be. Think tanks exist only as a label that is adopted or rejected for political, economic and social reasons (Medvetz 2012). This has provided think tanks with a great deal of flexibility in their engagement with their environment. They can play different roles in relation to research and its communication, depending on the contexts they face, the issues being addressed, and their own circumstances.

This diversity offers an advantage to think tanks that universities, by and large governed by similar rules across the world and which have emerged under similar patterns, cannot (and should not) exploit.

As a working distinction, though, I draw a dynamic and porous line between think tanks and research centres. The latter I understand to have an academic focus, either because of their location within an academic institution and the academic field or because of the purpose of the organisation.

Research excellence

This diversity presents an obvious challenge: what is the point of searching for a single measure of excellence when the interpretation of this concept is likely to be equally diverse. In developing countries, in particular, where think tanks fulfil roles that other weak institutions fail to (for instance, academia, policy-making bodies, civil society or the media) we would have to consider how excellence is defined by these other institutions, as well.

Can we compare excellence between think tanks undertaking teaching functions that universities fail to deliver, think tanks taking on public education and mass communication campaigns in the absence of credible news media or think tanks that provide policy analysis support to line ministries through consultancy or formal partnership agreements in light of limited policy analysis capacity within the civil service?

This diversity also opens the door to a common critique: think tanks do not care about research excellence, but only about their influence and their sustainability. This is what drives them and their choice of business model, their research agenda and communications strategies. This puts into question the legitimacy of their influence and the means they follow to achieve it. But, is it true that they do not care about excellence?

No think tank director would accept this. Short of asking them, one way to attempt to answer this question is to consider the way in which excellence is perceived by different types of think tanks; acknowledging that the types I will use are simply for illustration purposes and are gross oversimplifications, given the rich diversity mentioned above.

I draw on engagements with think tanks since 2010 through interviews, discussions and advisory work conducted as part of On Think Tanks to develop these perceptions. I also took advantage of the third Think Tank Initiative Exchange, held in Bangkok on 12–14 November 2018 and the third OTT Conference, held in Geneva on 4–7 February 2019 on the subject of public engagement, to discuss issues.

Think tanks with an academic origin or approach, for example, tend to perceive themselves as members of the academic community and are therefore bound to the same criteria of excellence as a research centre. This importance is illustrated in their choice of writing styles, the types of publications they prefer and the criteria they use to judge their performance: including publishing in academic journals, participating in academic conferences and staffing research positions with PhDs (as a proxy for an academic qualification). This is relevant for think tanks such as Grupo de Análisis para el Desarrollo (GRADE) in Peru and African Population and Health Research Center (APHRC) in Kenya. In evaluation terms, they are mostly concerned with the relationship between inputs (e.g. number of PhDs among their research staff) and outputs (e.g. number of publications of academic quality).

We could describe this as objectively verifiable excellence or what is traditionally recognised as an academic measure of quality. It is objective because there is little need to contextualise the indicators used.

Policy-driven think tanks, which would be comparable to the Anglo-American think tank model that is in most people's minds, but is far less common in developing countries, are far less concerned with academic credentials of excellence and instead seek confirmation that their research is relevant, timely and actionable. This is relevant for think tanks such as the Centre for Policy Analysis (CEPA) in Sri Lanka, Centre for the Study of the Economies of Africa (CSEA) in Nigeria or Grupo Faro in Ecuador. In other words, usefulness is included among the criteria of excellence. In evaluation terms, their focus shifts to the relationship between outputs and outcomes.

Think tanks with a strong membership base or close associations with other civil society groups such as workers' unions, business associations, political parties, grassroots or NGO networks would likely worry about the usefulness and ideological alignment of their research to that specific group. This is, coincidentally, also relevant for think tanks that depend on short-term consultancies from the government, the private sector or aid agencies. They are equally concerned about the alignment of their business models and their outputs to the interests and needs of their audiences.

Therefore, depending on how close they are to different communities (i.e. with academia, with politics and with civil society) and the nature of that relationship, think tanks assess research excellence differently. In consequence, we could argue that the only reason why academic think tanks worry about the robustness of their research methods or the verifiable excellence of their evidence is because that is the kind of thing that their main audiences, other researchers, would care about.

In other words, all think tanks search for is credibility within the communities they belong to or the communities they seek to influence.

Is the quality of the evidence produced by think tanks instrumental in awarding credibility?

The literature suggests that the quality of the research is not instrumental in awarding credibility, and therefore influence. Which does not mean that the quality of the research does not affect the quality of the advice and therefore the outcome of the decision made on the basis of said advice.

Fred Carden's often cited book, *Knowledge to Policy: Making the Most of Development Research*, does not consider the quality of the evidence used – in none of the 23 case studies included in the study (Carden 2009). The explanatory factors are mostly contextual and refer to the demand for evidence. In John Young and Julius Court's review of 50 case studies of policies informed by research, the quality of evidence is addressed only through the lens of the credibility of the evidence presented to policy-makers. 'Relevance – substantive and operational – clearly matters, but does the quality of the research? Although it is difficult to make a comment about the quality of the research in all the cases, the issue of credibility does come out as central' (Young and Court 2003: 16).

This study was one of the first to acknowledge the importance of considering different types of research and adopting a relatively loose definition 'from basic experimentation and social science research to policy analysis and action research' (Young and Court 2003: 9). Thus, the authors are unable to establish if the objectively verifiable excellence of research has any bearing on whether it is used or not.

Credibility, they argue, is far more important. Unlike the actual quality of the evidence, credibility does have a clear effect on its potential to inform policy. Credible think tanks and researchers gain access to decision-making space; credible evidence is used in the drafting of legislation; and credible policy arguments are adopted by policy-makers.

How is credibility gained?

Is credibility objectively or subjectively constructed?

Credibility is not a condition that can be objectively determined. Instead it is a construct of the interaction between researchers and think tanks with multiple actors and factors, over time, which provides a shared statement of their expertise and trustworthiness (Baertl 2018).

There are several characteristics of the research process that think tanks can control to some extent, including the quality of the data collected; the methods used to gather, store and analyse it; the quality of the writing; the design and publication of reports, etc. Some are more easily confirmed than others. Data quality may only be confirmed after a careful review or through replication studies. In contrast, the clarity in writing is something that any reader may assess on his or her own. However, even this is somewhat subjective; what may be clear to one reader may be impossible to comprehend to another.

In fact, the main factors affecting credibility are subjective and are subject to the judgements of think tanks' audiences: these may be other researchers, policy-makers, expert or epistemic communities, the general public, etc. Andrea Baertl's study on credibility identifies several factors (see Table 1 below), which offer think tank audiences different signals about the organisation, its researchers and its research excellence (Baertl 2018).

The factors mentioned in this overview offer signals of expertise and trustworthiness, the key components of credibility. These signals

Table 1: Factors determining credibility

Factor	Definition	Signals
Networks	Connections, alliances and affiliations that an organisation and its staff and board have.	Expertise Trustworthiness
Past impact	Any effect that a policy research centre has had on policy, practice, media, or academia.	Expertise
Intellectual independence and autonomy	Independence on deciding their research agenda, methods and actions an organisation undertakes.	Trustworthiness
Transparency	Publicly disclosing funding sources, agenda, affiliations, partnerships and conflicts of interests.	Trustworthiness
Credentials and authority	Collected expertise and qualifications that a think tank and its staff have.	Expertise
Communications and visibility	How and how often the think tank communicates with its stakeholders.	Trustworthiness
Research quality	Following research guidelines to produce policy relevant research in which the quality is assured.	Expertise Trustworthiness
Ideology and values	Ideology and values are the set of ideas and values that guide an individual or organisation.	Trustworthiness
Current context	The current setting in which a think tank and its stakeholders are immersed.	Frames the assessment and gives prominence to certain factors

Source: Baertl (2018)

are subjective assessments which are made with limited information, or because of the limited information that audiences have about the organisations, the researchers and their work.

For different think tanks, and depending on specific circumstances, these factors will have varying effects on their credibility. For example, the audiences of academic think tanks may probably pay greater attention to research quality itself, although access to research from an academic think tank is still likely to be mediated by the networks it belongs to and the reputations of the individual researchers. But how likely are they to review and attempt to replicate every research output published by the think tank – or are they more likely to rely on other signals? Did they use data sources that have been used in previous studies? Was it published in an academic journal? Who are the authors and where did they study?

Policy-focused think tanks will probably find that past impact and their values or ideology carry greater weight among politicians, who will no doubt be reassured by the ideological agreement with the premises of the research and the reputation of the researchers. The media will be particularly interested in their communications and visibility and the clarity and consistency of the message.

Ideology is an interesting factor. It can simultaneously confer credibility to a think tank in a community that shares its value and strip it of credibility in a community that doesn't. Andrew Rich's 2004 study of think tanks' visibility and influence in the United States demonstrated how credibility is in the eye of the beholder: when the Democratic Party controlled Congress, the most required think tank by congressional committees was the Brookings Institution; when the Republicans gained control, the Heritage Foundation gained the top spot.

At first glance, the robustness of the research methods used does not play a leading role in the assessment of a think tank's credibility and, therefore, its potential to inform policy.

This is true at different levels. For instance, Walter Flores (2018) found out that there is an inverse relationship between the level of academic complexity of the research methods and the levels of community engagement and responsiveness from the authorities. Figure 1 shows how the Center for the Study of Equity and Governance in Health Systems, in Guatemala, shifted its research methods over time. When it relaxed its research excellence criteria, it found greater engagement from communities and responsiveness from authorities. Flores concludes that:

> In contrast to theories of change that posit that more rigorous evidence will have a greater influence on officials, we have found the opposite to be true. A decade of implementing interventions to try to influence local and regional authorities has taught us that academic rigor itself

Figure 1: Academic complexity versus engagement and responsiveness

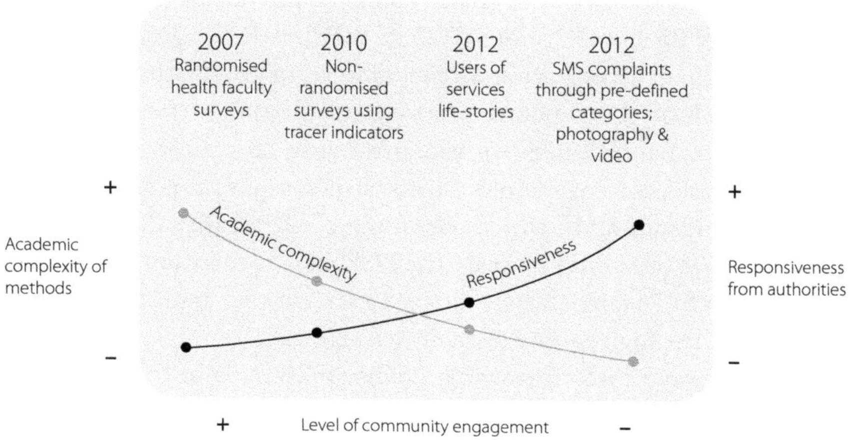

Source: Flores (2018)

> is not a determinant of responsiveness. Rather, methods that involve communities in generating and presenting evidence, and that facilitate collective action in the process, are far more influential. The greater the level of community participation, the greater the potential to influence local and regional authorities. (Flores 2018: 12)

Does research quality not matter at all?

The factors put forward by Baertl, the reviews by Carden, and Young and Court, and the case study presented by Flores suggest that the objectively verifiable quality of research does not play a significant role in the construction of credibility and therefore the influence that a think tank may have on policy decisions. But these accounts are snapshots of a moment in the lifetime of an organisation, a researcher or a single intervention.

These studies have not considered the long-term dynamics of credibility and how it is gained and lost. When we look at think tanks' efforts to influence policy over time, the objectively verifiable quality of research would take on a greater, albeit limited, role. For example, the Institute for Fiscal Studies (IFS) in the United Kingdom (UK) has built, over time, a reputation as the 'umpire' of the British economic debate. Much of this reputation is sustained by its accurate analysis of the budget, which it delivers, year after year, on budget day. Shoddy research would not have allowed it to build a reputation as a credible source of evidence and opinion. However, a BBC Reality Check article which asks: 'Why should we trust the IFS?' fails to mention the quality of its research. The article lists: no party affiliations, multiple funding sources and the high calibre of their researchers.

Another way in which research quality matters is in the quality of the advice it informs.

Opportunities for research and communication

These insights into how think tanks assess their credibility and the relatively low (but not negligible) importance that objective assessments of research excellence have on whether research findings are used or not, present several opportunities for effective communication that some think tanks have been able to exploit. These approaches go beyond post-research communication (they are embedded in everything the organisation does) and maximise the engagement of the think tank with their audiences (maximising the depth and length of such engagement).

In presenting the following approaches to research and communication, I wish to highlight a common ground with the standards of rigour expected from academic research, the implication being that some of these approaches could therefore be adopted without compromising objectively verifiable quality.

In addition, they would enable the assessment of the credibility of research to go beyond the outward facing factors identified by Baertl and allow a more nuanced approach based on more objective criteria of quality. This is possible because all of these approaches have a common, secondary objective: to build trust between think tanks and their various audiences or publics. In doing so, think tanks are able to establish a relationship that can look beyond subjective notions of credibility (because there is trust already) and focus on more objective assessments of quality.

Organisation-wide branding

John Schwartz, director of the communications firm, Soapbox, has written about the role of branding for research. Soapbox works with think tanks and universities, helping them to communicate recommendations, research findings and even their research practice, itself. In a recent series of articles based on their experience with the Stockholm Environment Institute (SEI), Schwartz (2018) argues that brands help research organisations:

- Become the organisations they aspire to be;
- To own a piece of intellectual and cultural territory; and
- Produce the right kinds of communication for the right audiences.

In an environment saturated by information, research centres need to find new ways to stand out as the most credible sources. This means that every aspect of the organisation's work, from its office space to its research, publications, events and social media, is an opportunity to reinforce its expertise and trustworthiness.

In practice, research centres have left behind the project-specific and ad-hoc communication efforts of the past to instead develop coherent organisation-wide communication strategies. These encourage and nurture a relationship with their audiences which goes beyond specific individuals, research findings or recommendations and encompasses a broader range of services and experiences which maximise an increasingly nuanced engagement.

Public engagement rather than elite influencing

Think tanks are increasingly concerned about public engagement rather than direct policy influence. This is a result of two emerging ideas: first, credibility matters and, second, the general public is an increasingly important player; both in awarding credibility and influencing policy. Nick Pearce, from the University of Bath and former director of the Institute for Public Policy Research, reflected at an event in early 2018 that, post-Brexit referendum, think tanks in the UK have recognised the importance that the public plays in the outcome of policy debates and policy decisions.

In liberal democracies where politics have taken a more polarised nature, think tanks have turned towards the public as a vehicle to reclaim more moderate, evidence-informed, debates. Think tanks in contexts where the civic space is rapidly, and violently, shrinking have adopted communication strategies increasingly aimed at boosting their credibility with the general public. At the same event, Sonja Stojanovic Gajic, from the Belgrade Centre for Security Studies, agreed that this applies to several think tanks in the Western Balkans.

However, meaningfully reaching the public demands a different approach to reaching the political, economic or social elites to which think tanks have been accustomed. The public's interest and understanding of the issues is highly heterogeneous. Also, the means by which they have arrived at that understanding or the opinions they hold may be different to those preferred by think tanks and the scientific community more broadly. There are no obvious policy recommendations for them to act on. And, their knowledge of, or their opinion of, think tanks themselves is limited – with obvious consequences on their credibility. Recent polls in the US (Hashemi & Muller 2018a) and in Britain (Hashemi & Muller 2018b) show that the majority of the public do not know what think tanks are or what they do. Why would they trust them?

This requires an approach that combines audience segmentation, developing narratives and different levels of engagement. In practice, this means that policy research centres are increasingly investing in editorial capacity (to write for different audiences), paying greater attention to the development of comprehensive narratives and producing multiple communication outputs which are disseminated through multiple communication channels (to facilitate different types and levels of engagement from these different audiences).

Figure 2: Pyramid of engagement

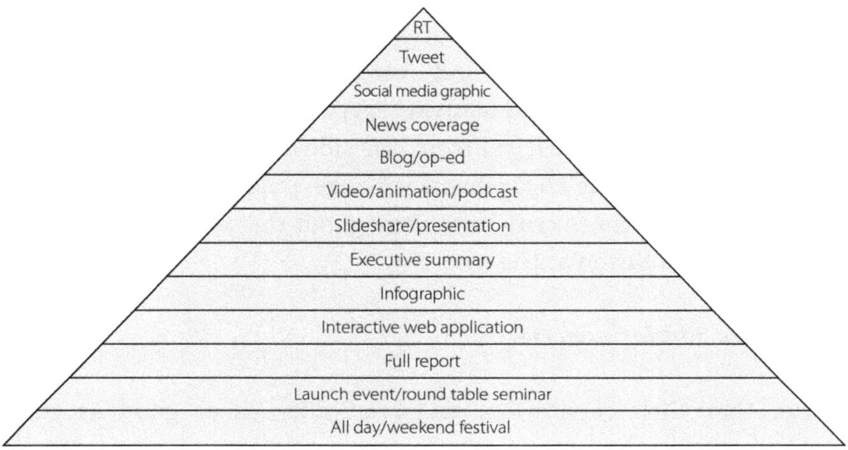

Source: WonkComms: https://wonkcomms.net/2013/10/17/videos-and-slides-wonkcomms-in-the-north/

Richard Darlington argues for a pyramid of engagement for research (Darlington 2013). In Figure 2, Darlington presents this as an alternative to what he calls the 'submarine strategy': when researchers go deep under water for long periods of time while they study an issue in full and until their work is finally published (Darlington 2017). This approach fails to recognise the progressive nature of change, and it assumes that the robustness of the evidence, when it is published, will be sufficient to sway opinions.

Greater engagement offers multiple opportunities to address entrenched beliefs based on incorrect evidence or in spite of the existing evidence. Over time and through different engagement activities, think tanks may build trust – a key component of credibility – and progressively sway opinions. They may also help the public, and particularly those among them who distrust the scientific method, to better understand the research process, the evidence it produces and its implications.

Again, this is in contrast to the traditional, one-way approach to research or scientific communication, which assumes that the public holds opinions contrary to what the evidence suggests, because the public has not had access to that evidence.

Greater engagement also exposes think tanks to sustained scrutiny from multiple audiences and over a longer period of time. This is meaningful in a polarised context: every policy recommendation is likely to be met with criticism from one side or the other of the aisle. Therefore, continuous engagement with the public can help identify, raise and address those criticisms throughout the process, thus avoiding a head-on collision at the end.

Conveners, not influencers

Aware that their reputation, and credibility, is as good as their last growth prediction or policy recommendation, think tanks are increasingly turning their attention to creating spaces to convene, rather than actively and overtly attempt to influence, policy actors. One of Chatham House's most recent approaches to communication is the use of simulation exercises that present them with the opportunity to offer their evidence and advice in a safe environment and in a useful way. According to its head of communications, Keith Burnett, this approach also allows the centre to incorporate evidence from multiple sources, thus emphasising their position as trusted conveners (Burnett 2019).

Think tanks across Latin America have turned their attention to electoral processes (Echt 2015; Echt and Ball 2018). While some of these efforts aim to present clear policy recommendations and seek to directly influence the agendas of future governments, most, in fact, have focused their efforts on informing the debate and, on occasion, staging the technical and presidential debates themselves. They have been more successful when their efforts have involved multiple organisations and voices.

This presents them as party neutral and impartial, knowledgeable and well connected; in other words, as credible, and it promotes greater engagement between the research and the researchers and their audiences.

Windows of opportunity

The focus on elections stems from the recognition that most think tanks' resources are limited. Sustaining a prolific research and communication production all year long is possible only for a handful of think tanks. Most think tanks are small, resource-strapped and only occasionally influential. Furthermore, their funding is mostly project based, which makes it difficult to focus on a single issue in the long term.

How, then, can they build the credibility they need to be influential and offer their multiple audiences the appropriate opportunities for engagement? An effective strategy is to target predictable policy windows with research and engagement campaigns.

For example, the British IFS has become the most credible source of analysis of the budget on budget day (Akam 2016). Arguably, the quality of their analysis is greatly responsible for this. But of similar importance is the manner in which they have turned the entire organisation on this brief, albeit important, window of opportunity.

This approach can have long-lasting effects. Elections are also excellent windows to consider. The Center for the Implementation of Public Policies for Equity and Growth (CIPPEC), in Argentina, has successfully targeted several presidential elections to take centre stage in the policy debates that dominate the news media. On its third attempt, CIPPEC had positioned itself to inform and staff President Mauricio Macri's administration. Its policy recommendations were presented at the exact moment when the future government was in search of ideas and people (Echt and Ball 2018). A year earlier, the same ideas and people would not have attracted the same level of interest. As a consequence of this, the new government turned to CIPPEC in search of experts to join their ranks. This, in turn, makes it possible to better inform policy-making.

In Chile, the smaller and newer Espacio Público also used Chile's 2018 elections to establish its reputation on corruption. It launched a research-based campaign on party financing which succeeded in setting the agenda by taking advantage of the public's natural distrust of political parties and electoral financing. At any other moment in time, the complex nature of the subject would not have attracted the same level of support. The campaign has led to the establishment and launch of a regional network focused on research on anti-corruption policies.

A focus on transparency

The Transparify initiative, launched in 2014, has helped usher a new wave of efforts from think tanks, their funders and the media to promote the financial transparency of public policy research. Although Transparify only covers a small number of think tanks in the world, many have adopted their approach and have even requested a bespoke review. This effort to open up presents think tanks with an opportunity to engage with an audience that is increasingly incredulous about the credentials that experts claim for themselves.

Implications

In this chapter I have attempted to explore the effect that a focus on credibility, rather than objectively verifiable quality, has had on think tanks' strategies. Regardless of their business models, think tanks have elected to adopt research and communication strategies that effectively maximise the depth and length of engagement with their different publics and attempt to draw attention to the factors that help build their credibility. In other words, think tanks are segmenting their audiences to establish a closer relationship with individual groups.

Somewhat counterintuitively, these trust-building strategies present an opportunity and represent an effort to move away from a notion of credibility based on perception (e.g. networks, visibility, past impact, etc.) towards one based on a more rigorous assessment of quality. That is, to establish an evidence base for expertise and trust that may be objectively verified by the members of the spaces that think tanks now share with their audiences.

These approaches have important implications for university research centres and researchers. First, they demand a greater commitment to public engagement that most currently have. In particular, the *public* here must be understood to include not just the student body and alumni, but also the individuals and institutions that belong to their polity. Naturally, this involves an effort by the entire organisation and not only communicators. The experience in Guatemala illustrates this. Researchers have had to adapt their research methods and involve communicators in their design. Furthermore, the organisation has had to adapt and encourage innovation in this field.

Second, these approaches reject the claims of influence and rankings. What matters is not the number of citations (which may or may not be based on a nuanced assessment of quality) but the quality of the engagement of key audiences with the research, the researchers and the organisation.

Third, the research output is no longer the bridge between producers and users of evidence and knowledge. The focus must be on the relationship between them, and this relationship is fundamentally held by individuals and their practice.

Finally, across all approaches, one finds a greater commitment to disclose the role of the organisation and the way in which evidence and advice is formulated. Greater transparency (financial and otherwise) can significantly contribute to the development of stronger relationships and a more nuanced assessment of quality.

Conclusion

The diversity of think tank formation and development has created fertile ground for innovation with respect to how they communicate evidence and advice. Their emphasis on non-academic impact demands that they pay attention to how multiple audiences perceive them and their work. While objectively verifiable assessments of the quality of their research are important, subjective factors make a greater contribution to the credibility of think tanks and their research.

In their search for credibility they have adopted rather successful approaches to communicating evidence that are compatible with research centres which, by their nature, place greater emphasis on objectively verifiable indicators of research excellence. These approaches, in fact, make it possible to develop new relationships that facilitate a deeper and longer engagement, which has the effect of refocusing the assessment of credibility from subjective to objective criteria.

To establish and maintain these relationships, however, university research centres will have to usher in important changes to the way they are managed, funded, the way they undertake research, and the strategies they use to communicate. This does not demand a change in their missions but an acceptance that they may be better served by adopting a more nuanced understanding of research quality and impact.

References

Akam S (2016, 15 March) The British umpire: How the IFS became the most influential voice in the economic debate. *The Guardian*. Retrieved June 26, 2018. https://www.theguardian.com/business/2016/mar/15/british-umpire-how-institute-fiscal-studies-became-most-influential-voice-in-uk-economic-debate

Baertl A (2018) De-constructing credibility: Factors that affect a think tank's credibility. *OTT Working Paper Series 4*. Lima and Bath: OTT

Belletini O (2007) El papel de los centros de investigacion de politica publica en las reformas publicas implementadas en America Latina. In: A Garcé and DE Abelson (eds) *Think Tanks y politicas publics en Latinoamerica: Dinamicas globales y realidades regionales*. Buenos Aires: IDRC/Konrad Adenauer Stiftung/Prometeo libros

Burnett K (2019) Immersive experiences and simulations are helping think tanks adapt to the changing policy landscape. *OTT Annual Review 2018: Public Engagement*. OTT. https://onthinktanks.org/wp-content/uploads/2019/02/OnThinkTanks_OTT_AnnualReview_2018.pdf

Carden F (2009) *Knowledge to Policy: Making the Most of Development Research*. Ottawa: IDRC & SAGE

Darlington R (2013) The pyramid of engagement. *WonkComms in the North*. Retrieved on 27 June 2018. https://www.youtube.com/watch?v=UWspTvK5Ki8

Darlington R (2017) Defying gravity: Why the submarine strategy drags you down. *Wonkcomms*. Retrieved on 27 June 2018. https://wonkcomms.net/2017/08/16/defying-gravity-why-the-submarine-strategy-drags-you-down/

Echt L (2015) Think tanks and elections: A series on experiences from around the world. *On Think Tanks*. Retrieved on 27 June 2018. https://onthinktanks.org/articles/think-tanks-and-elections-a-series-on-experiences-from-around-the-world/

Echt L and Ball L (eds) (2018) Think tanks: Why and how to support elections. *On Think Tanks*. https://onthinktanks.org/wp-content/uploads/2018/11/OTT_ThinkTanksandElections_November2018.pdf

Flores W (2018) How Can Evidence Bolster Citizen Action? Learning and Adapting for Accountable Public Health in Guatemala. *Accountability Note 2*. Accountability Research Center

Hashemi T and Muller A (2018a) *Forging the Think Tank Narrative (USA): Credible but not Effective Communicators*. We are Flint report. Retrieved on 25 June 2018. https://weareflint.co.uk/forging-the-think-tank-narrative-perceptions-usa/

Hashemi T and Muller A (2018b). *Forging the Think Tank Narrative (UK): Credible but not Effective Communicators*. We are Flint report. Retrieved on 25 June 2018. https://weareflint.co.uk/forging-the-think-tank-narrative-uk

Mendizabal E (2012) Think tanks en Latinoamérica: ¿Qué son y qué los mueve? *Foreign Affairs: Latinoamérica* 12(4): 68–76

Mendizabal E (2013, 28 January) Think tanks in Latin America: What are they and what drives them? *On Think Tanks*. Retrieved 25 June 2018. onthinktanks.org

Mendizabal E (ed.) (2016) Think tanks in China. *OTTSeries*. https://onthinktanks.org/series/think-tanks-in-china/

Mendizabal E, Datta A and Jones N (2010) Think tanks and the rise of the knowledge economy. In: A Garcé and E Uña (eds) *Think Tanks and Public Policies in Latin America*. Fundación Siena/CIPPEC

Medvetz T (2012) *Think Tanks in America*. Chicago: The University of Chicago Press

Puryear JM (1994) *Thinking Politics: Intellectals and Democracy in Chile, 1973–1988*. Baltimore: The Johns Hopkins University Press

Rich A (2004) *Think Tanks, Public Policy and the Politics of Expertise*. Cambridge: Cambridge University Press

Stone D and Denham A (2004) *Think Tank Traditions: Policy Research and the Politics of Ideas*. Manchester: Manchester University Press

Schwartz J (2018, 27 March) SEI's communications revolution: Part 1 – rebranding. *On Think Tanks*. Retrieved on 26 June 2018. https://onthinktanks.org/articles/seis-communications-revolution-part-1-rebranding/

Villanueva A and Mendizabal E (2016) *Impacto social de la investigación: Aproximaciones, desafíos y experiencias internacionales de evaluación*. Research Report, Pontificia Universidad Católica del Perú, Peru

Young Y and Court J (2003) Bridging research and policy: Insights from 50 case studies. *Working Paper 213*. London: ODI

PART

3

Striving for solutions

CHAPTER 12

Exploring research evaluation from a sustainable development perspective

Diego Chavarro

Introduction

Evaluation studies produce 'value judgements about the quality, worth, or value of intervention programmes' (Mouton 2014: 64). These value judgements usually rely on standards, theories and ideals as points of reference against which the assessment is performed. They are lenses through which an intervention is analysed, and therefore determine what is valued and how. For instance, an evaluator using a conventional economics perspective will likely value policy interventions for their contribution to solving market failures (Dollery & Worthington 1996), whereas a constructivist will likely value more the process of inclusion and debate in policy co-construction (Guba & Lincoln 1989). The perspective chosen has implications for the type of analysis performed, the unit of analysis chosen, the assessment criteria used and the methodology for the evaluation. Therefore, discussing the theories underpinning evaluations helps understand their rationales, usefulness and limitations. This is relevant for policy-making because, when some evaluation frameworks become dominant or popular, they are used indiscriminately for purposes that do not match their aims.

Research evaluations are based on theories about the value of scientific knowledge production (Molas-Gallart and Ràfols 2018). Countries such as Australia, UK, Brazil and Colombia have developed national Research Evaluation Systems - RES (Chavarro 2017), which can be understood as 'organised sets of procedures for assessing the merits of research undertaken in publicly funded organisations that are implemented on a regular basis, usually by state or state-delegated agencies' (Whitley & Gläser 2007: 6). RES can have different theoretical underpinnings, but one that is common to many of them is the sociology of science perspective, which values research for its 'scientific impact' and 'scientific quality' (Chavarro et al. 2018). Frequently, RES use quantitative indicators such as citation and bibliographic output counts to measure such concepts. These will be referred to as production and citation indicators. The popularity of these indicators has given scientometrics, or the quantitative study of science (Wouters 1999), a preponderant role in RES.

However, the evaluation of scientific knowledge production based only on 'intrinsic', 'scientific' criteria is being challenged by alternative perspectives that value knowledge production for 'extra-scientific' criteria, such as its impact on society, institutions and environment (Orozco et al. 2007). Sustainable development, which can be understood as a balance between economic, environmental and social development (Gallopín 2001), is one of those extra-scientific criteria by which research can be evaluated. This perspective challenges RES based on the sociology of science because sustainability demands the social accountability of knowledge construction.

In this chapter I focus on the examination of some of the theoretical underpinnings of conventional RES, focusing on RES that use production and citation indicators to produce rankings, discussing their limitations in capturing the features of knowledge construction in a context of sustainable development. I also discuss, from a policy-making perspective, some of the reasons why transforming national RES faces resistance, suggesting alternatives that can be explored by research councils and other policy organisations wanting to develop evaluations in the context of sustainability.

Some foundations of scientometric indicators for research evaluation

Scientometrics is a discipline devoted to the quantitative study of science, technology and innovation (ST&I). One of its main areas of research is the development of indicators for research evaluation. Scientometric research evaluation focuses mainly on the assessment of two properties[1] of scientific knowledge: production and quality (Molas-Gallart and Ràfols 2018). Usually, these properties are expressed as some measure of quantity of scientific outputs (articles, books, patents) and citations to scientific literature respectively.

For Merton, the function of science is the production of certified knowledge (Merton 1973)[2]. This knowledge is communicated in journals that are recognised by scientific communities as the gatekeepers of quality research, overseen by a system of peer reviewers who evaluate the soundness of scientific contributions based on disciplinary standards. For Merton, the production of certified knowledge is considered a social value in itself, so considerations about the social utility of that knowledge are extra-scientific and not necessary in order to justify funding decisions.

Derek de Solla Price and Eugene Garfield were fundamental in diffusing Merton's sociological description of scientific norms and Mertonian ideas of scientific production. Price set the foundations of scientometrics and Garfield operationalised these ideas and analyses through the development of the Science and Social Sciences Citation Indices and his invention of the citation indicator (Wouters 1999; Godin 2006a). Although initially the intention of Garfield was not to produce indicators for rankings, rankings of researchers, journals, organisations and countries became the main use of production and citation indicators (Chavarro 2017). This use fits the Mertonian conception of science because he depicted the scientific system as a hierarchical structure. In this hierarchical structure, some scientists receive more recognition than others due to their experience and the significance of their contributions to science, which brings them reputation and scientific authority. In the end, RES based on production and citation indicators reproduce this emphasis on reputation, because many of them aim to distribute funds and public recognition to individuals and organisations in a competitive manner.

The above understanding reproduced by production and citation indicators embedded in RES[3] fits into a certain framing of research policy, referred to as frame 1 ST&I policy (Schot and Steinmueller 2016) or linear model (Godin 2006b). This frame is portrayed as a sequential process in which research activities are an input for technology development, and technology development is the engine for economic growth. In this conception, the role of the state is to fund researchers and innovators, leaving the decision on what to research to them because there is an implicit assumption that all science and innovation brings positive outcomes in terms of economic growth (Schot and Steinmueller 2016). In this sense, reputational indicators such as the ones discussed earlier fit a policy framework that conceives science as a means for economic growth. The outcomes of many RES, especially rankings of scientists and organisations, are then the recognition by the state to those who contribute the most to economic growth.

To summarise, production and citation indicators are frequently used in a way that considers the production of certified knowledge an aim in itself. This conception of science rewards individuals and organisations who excel at publishing their research, creating a system in which scientists compete for public recognition and reputation. Questions about the utility of science are not covered by this conception because it is assumed that all knowledge is beneficial to society and scientific development produces positive (economic) outcomes. In the next section, I will present the limitations of this conception to address new societal demands essential to sustainable development.

Sustainable development challenges traditional scientometric indicators

The concept of sustainable development does not have a single definition. However, the historians of the concept place its origins in the environmental movement and in environmental economics (Meadows et al. 1972). Although there are different definitions, it can be concluded that sustainable development seeks social, environmental and economic balance (Chavarro et al. 2017). This balance, however, is a 'dynamic equilibrium' (Gallopín 2001) because the three systems are in constant renovation.

In current public policy, sustainable development has become a framework of wide acceptance, mainly due to the thrust of multilateral organisations such as the United Nations (UN) and the Organisation for Economic Co-operation and Development (OECD). These organisations have managed to convene several countries over time and around joint programmes of cooperation. One of these programmes is the 2030 Sustainable Development Agenda, which seeks to achieve 17 social, economic and environmental goals known as the Sustainable Development Goals (SDGs) (see UN 2015).

Within this sustainable development agenda, science is not considered only a source of new certified knowledge or the engine of economic growth, but also a contributor to solving social and environmental issues. The Scientific Advisory Board (SAB) of the UN Secretary General has acknowledged the 'crucial role of science for sustainable development' (SAB 2014), and published a report discussing some of the ways in which science is related to the achievement of SDGs. According to this Board, science has the following roles:[4]

- Provides the basis to identify and face global challenges;
- Offers a mechanism to cross the 'national, cultural and mental barriers', which is necessary to work collaboratively in the challenges of sustainable development;
- Through scientific literacy, provides education and helps build the capacity to use science to solve everyday problems;
- Can strengthen democratic practices if it is treated as a public good;
- Thanks to its ability to integrate knowledge from different disciplines, it helps to face challenges that are interdependent (for example poverty, economic growth, clean water and clean energy);
- Provides evidence to formulate ST&I public policy and to inter-relate other public policies;

- Helps to monitor progress in the different sustainable development objectives; and
- Education with a strong scientific component prepares societies to respond creatively to the challenges presented to them.

From the list above it can be derived that the role of science in sustainable development differs from its role according to the sociology of science. The main difference is that in sustainable development the production of new knowledge is not valued on its own, but on its relevance to the environmental, social and economic challenges of the world.

With regard to policy framing, sustainable development is not adequately represented by the linear model or even by the national systems of innovation model. For this reason, scholars are developing specific policy frameworks to conceptualise the role of science in sustainable development. One of these frameworks is the Transformative Innovation Policy (TIP), which proposes that to transition towards sustainable development there is a need to exert profound changes in the systems of consumption and provision of goods, as well as in culture and economy (socio-technical systems) (Schot and Steinmueller 2016). TIP understands ST&I as cross-cutting to all SDGs, and as a means to achieve socio-technical changes towards sustainability. Based on the above points, I present a list of four principles, understanding principles as properties that research evaluation should consider when appraising science in a context of sustainable development.[5] I also summarise some of the key challenges posed by sustainable development to traditional research evaluation:

- *Transformation*: Sustainable development requires transformation of socio-technical systems, not only the production of new knowledge. Therefore, sustainable development demands involvement of scientists in solving environmental, economic and social challenges. Increasingly, research councils are requiring scientists to show the social impact of their research; see for instance, the UK's Research Excellence Framework or the NSF's broader impact criteria for research proposals, which show a movement towards this direction. However, social impact is only a part of sustainable development, which requires 'transformative impact'. This implies the development of evaluation frameworks to account also for the environmental and economic impacts of research, as well as their inter-relationships.

- *Collaboration*: Sustainable development is about collaboration, not competition as promoted by RES focused on ranking systems. In the context of sustainable development, collaboration includes international and intra-national collaboration, interdisciplinarity and inter-sectorial cooperation.
- *Directionality*: Sustainable development requires directionality, in the sense that not all scientific production or innovation will bring positive sustainable effects. For instance, from a sustainability perspective, research on chemical weapons with the aim to produce them on a large scale is not desirable, even if it is highly cited.
- *Participation*: Sustainable development requires the participation of government, citizens and entrepreneurs, in addition to scientists (which is not favoured by the sociology of science or linear model approaches).

The list is based on my interpretation of the role of science from a sustainable development point of view, in comparison to the sociology of science or linear model conceptions. It is presented, then, as a list for discussion. However, even in this preliminary elaboration, it can be seen that traditional ST&I indicators do not account for the properties of sustainability. Despite their unsuitability, similar indicators are being used to measure progress on sustainable development. For instance, the UN elaborated a set of indicators to measure progress on goals related to sustainable innovation; the ones proposed to measure progress on research and innovation capacity are research and development (R&D) expenditure and number of researchers per inhabitants.[6] These indicators are based on a linear model perspective, which assumes that more is better, but does not address issues such as on what subjects the researchers are working, what research is being funded, who are being funded, or how interdisciplinary and transformative the research is. Therefore, the use of these indicators as proxies for sustainability is a case of 'streetlight effect' or 'drunkards search', meaning that their use can only be justified because they are the indicators at hand (Molas-Gallart and Ràfols 2018). In this way, cases such as the UN's choice of traditional ST&I indicators as proxies for sustainable innovation fail to capture the sustainability concept they intend to measure. A question arises as to what can be done, then, to include sustainability concerns in research evaluation. In the next section, I suggest some ideas to consider in the design of indicators to address this issue.

Ideas for research evaluation systems (RES) in the context of sustainable development

RES are an important component of ST&I policy, because they help to steer research in desired ways through recognition and funding (Whitley & Gläser 2007). Although it is not clear how effective RES are in actually shaping research agendas and setting priorities towards desired goals (Rijcke et al. 2016), they make visible what is valued by research councils and funding agencies. For this reason, if sustainable development is to be supported by a country, its RES needs to incorporate clear criteria in the direction of sustainability. Such criteria, however, are absent from many RES that continue to apply an evaluation model based on production and citation indicators. Here I provide some ideas on why this continues to happen and suggest points to be considered when designing research evaluations in the context of sustainable development, attending to the aforementioned principles. For this I use the case of Colombia, a country with an RES which is heavily based on production and citation indicators. I start by describing the Colombian RES of research groups, attempt to understand why the foundations of this RES have resisted structural modifications, and then suggest some of the points that could be included in it if the country is to promote science for sustainable development.

In Colombia, as well as in other countries such as Mexico, Brazil, Chile and Spain, RES based on production and citation indicators affect directly or indirectly the distribution of funds and recognition to researchers and organisations (Chavarro 2017). Colciencias manages two large research assessments, one for research groups (GrupLAC) and the other for journals (Publindex). Recently, Colciencias has also started to evaluate individual researchers. Groups are evaluated according to a quantitative index composed mainly of a weighting scheme applied to their bibliographic output. This RES underwent several changes from 2000 to 2015 (Nupia 2018). The first change was to introduce the calculation of a score to measure different types of bibliographic outputs according to quality criteria, mainly citations to research papers. Another change was the introduction of other outputs of importance to disciplines such as arts and architecture, for instance, evidence of concerts, performances, paintings, novels and blueprints. Other modifications include the criteria to endorse research groups, or weights given to different types of outputs, and recently the measurement model has included scores for non-bibliographic results such as spin-offs.

However, most of the changes introduced to the Colombian RES have not questioned profoundly the principles of production and quality under which it is built, and the indicators used. I queried experienced colleagues who have worked on the development of Colciencias' research evaluation model as to why they think its principles seem to be accepted, complementing their answers with my own experience and with literature search. My findings are summarised in the following list:

- *Stability*: since 2000 the criteria for assessing research have remained relatively stable, as well as the way to measure these properties (a quantitative model that produces a score);
- *Routine*: once established, the procedure of measuring research groups was codified in software. The software is run each year and this has become part of the organisations' routine with an allocated yearly budget;
- *Predictability*: experience has taught Colciencias how to deal with software errors and complaints from researchers and institutions; the software even offers simulations to predict the ranking of the research group before the actual evaluation is performed;
- *Co-construction*: the criteria have been debated between Colciencias and representatives of the academic community (Nupia 2018);

- *Flexibility*: even though the model is based on a formula, the discussions with the academic communities have allowed the incorporation of new products into the measurement;
- *Link with distribution of funds*: in public universities, scores obtained from scientific production indicators represent a salary increase for teachers. Also, the classification of a research group may determine its eligibility for funding;
- *Isomorphism*: the research evaluation model, which directly affects public universities, is being reproduced in private universities, some of which give economic incentives to researchers for their scientific production indicators; and
- *The model works*: citations and production counts are readily available and have become a standard. In comparison, other indicators are less developed in terms of their reliability and their interpretation is even less clear (e.g. altmetrics).

The current RES may have been able to allow for minor changes to the calculation of rankings, but the work that Colciencias is doing on the design of an ST&I policy for sustainable development (Colciencias 2018) has shown that to incentivise the contribution of ST&I to solving grand challenges, the current research evaluation model needs to be renewed. In order to include the principles of sustainable development in evaluation for sustainability, Colciencias' RES could:

- Incorporate ways of appraising scientific collaboration and the participation of diverse social groups in research activities (citizens, entrepreneurs, NGOs, etc.) – participation and collaboration principles;
- Increase efforts to use content analysis in evaluations of research outputs instead of citation counting only, because sustainable development research is directed towards specific goals. Assessing research related to the SDGs, for instance, requires identification of the specific subjects being researched. Semantic analysis can give relevant evidence to policy-makers to steer research funding and promotion – directionality principle;

- Prioritise the communication function of science over the use of research evaluation indicators for career development and individual or organisational reputation. This may imply unlinking research evaluation from direct funding of researchers and institutions, and instead allocating funds to subjects or topics of national or local interest, as well as problem-oriented research – transformation and directionality principles;
- Incorporate ways of appraising environmental and social impact in addition to scientific and economic impact – transformation principle;
- Reward novelty and relevance of contributions, not only accumulation (e.g. of citations) – transformation principle;
- Incentivise science as a public good, as opposed to science as a private endeavor. This may imply giving more weight to research that can be openly distributed, but also to research that uses or builds open infrastructures, etc. – collaboration and participation principles.
- Incentivise interdisciplinarity, because it is needed to address local issues – transformation principle; and
- Include non-traditional research outputs, such as technical manuals or other products that are difficult to codify in standard bibliographic outputs and can have transformative impacts – transformation principle.

However, based on conversations with colleagues in Colciencias and on my own experience in research evaluation, I find that there are different barriers to implementing radical changes. For instance, changing an established RES requires huge investments both in terms of funds and time, thorough discussion with academics and other stakeholders if the principle of participation is to be put into practice, and restructuring areas of Colciencias that are devoted to managing the current RES. Connected with this, a new RES requires creating internal capacity. This capacity refers not only to technical skills, but also to the suitability of the legal framework to accommodate a new research evaluation model, which could have an impact on the salaries of researchers and the distribution of funds. Also, some may see modifications to the current RES as a threat to 'scientific quality', given that, in the case of sustainability, the 'relevance' criterion is just as important. Some may even argue that the basic sciences would be disadvantaged by a 'utilitarian' understanding of knowledge production. Therefore, from a policy-maker's perspective, it is not an easy decision to radically change the current RES. However, some alternatives could be explored.

One alternative is to adjust the current quantitative model to include and give weight to some of the principles of sustainable development. This is the approach that has been followed in the past 15 years to introduce changes. Another alternative is to have two separate measurement models, one to award 'quality' and the other to award 'relevance'. This would require two research evaluation systems, which would be very costly and operationally demanding. A third one could be to have a multidimensional model, with one of the dimensions being 'quality' and the other being 'relevance'. The question here is how to weight the dimensions: which is the more important, quality or relevance? In summary, any option implies trade-offs between different valuations of science.

A recent attempt to bridge the 'quality' and 'relevance' gap, addressing the issues above, is the Research Quality Plus (RQ+) framework designed for the evaluation of research for development (Ofir et al. 2016). The idea behind RQ+ is that research quality is a multidimensional concept, which goes beyond scientific merit. The framework is highly customisable, and offers ways to include key influences that constrain research, the different dimensions of quality beyond citations (integrity, legitimacy, importance and positioning for use) and rubrics for assessing each component. By employing this framework, researchers at IDRC have found that research for development produced in the Global South outperforms research for development produced in the Global North, contradicting most of the studies of scientific production based on production and citation indicators. This shows that evaluations of research are dependent on how 'quality' and 'relevance' are defined.

Including sustainability principles such as transformation, directionality, collaboration and participation in the quality dimension of RQ+ and developing rubrics for assessing them is something that is allowed by the framework. How the inclusion of these principles would change quality assessments of research in the Global South in different disciplines is a question worth exploring. Although promising for evaluations of research in a context of sustainable development, novel frameworks for research evaluation have yet to be tested in countries such as Colombia that rely heavily on production and citation bibliographic indicators. There is, then, a great opportunity to conduct pilots to learn how concepts such as sustainable development can be included in research evaluation and the acceptance of, or resistance to, novel ways of research evaluation by research communities and other social groups.

Conclusion

The purpose of this chapter is to contribute to a better understanding and use of scientometric indicators and to help develop principles to guide the design of new indicators and research evaluation in the context of sustainable development policy. By examining the theories that underpin indicators' development and use, as suggested by Mollas-Gallart and Ràfols (2018), it was possible to see why some research evaluations fail to convey the properties that they try to measure. This was done by exploring some of the assumptions underpinning RES, based on scientometric indicators and comparing these assumptions to those of sustainable development. My argument is that the conventional scientometric indicators used by RES cannot evaluate research in the context of sustainable development, mainly because they are based on a theory of science that regards the production of certified knowledge as a social value in itself, whereas sustainable development values it in relation to its relevance for social, environmental and economic issues. For this reason, if policy-makers want to develop research evaluations to support sustainability, there is a need to understand this radical difference and design alternative indicators and evaluation frameworks that reflect sustainability more accurately. I have also suggested some changes that could help to produce more sensible research evaluations that meet their stated objectives. Basically, RES that want to better represent the concept of sustainability could include criteria to address transformation, collaboration, directionality and participation criteria, which are absent from conventional scientometric evaluation.

Despite the need for alternatives, transforming a RES is challenging for a research policy organisation. By using the case of Colombia as an example, this chapter provides a concrete account of how discussions on research evaluation materialise in the decisions that a research policy organisation must make and why changes, which appear to be relatively 'simple' from an academic point of view, are complex in practice: the resources devoted to the development of an RES, the time required to establishing it, the routines developed around its implementation, the regulations and funding linked to it, the human capacity needed in order to operate it, and potential criticisms are constraints that policy-makers face when taking the decision to embark on a new RES.

Despite the above constraints, evaluation frameworks such as RQ+ offer a way to test alternative understandings of research quality and incorporate new criteria, such as sustainability, in a way that bridges the gap between 'quality' and 'relevance'. Conducting evaluation pilots in countries such as Colombia will contribute to establishing the usefulness and limitations of these novel frameworks, and their complementarity to conventional research evaluation. Although it can be seen as a costly exercise, the benefits of experimenting and learning will outperform its cost, which is applying a conventional evaluation instrument that does not fit the new societal demands from science.

Acknowledgements

This chapter is based on my experience as an advisor to the Policy Evaluation and Design Unit of Colciencias, which is the organisation in charge of ST&I policy in Colombia. I have drawn also on conversations with colleagues and discussions in different workshops on the development of an ST&I policy for sustainable development. I am especially indebted to my colleagues who participated in the SGCI/IDRC workshop on 'Perspectives on Research Excellence'. My special thanks go to Johann Mouton, Erika Kraemer-Mbula and Matthew L. Wallace for their insightful comments.

Notes

1 I use the word 'property' in the sense pointed out by Molas-Gallart and Ràfols (2018), who state that indicators are ways to approximate the measurement of properties that are not directly observable. Quality, for instance, is a vague term to indicate a property that makes a scientific outcome more valuable than another. As quality cannot be observed directly, an indicator such as the number of citations can give an idea of the level of quality of that product, of course assuming that researchers cite others based on quality considerations. In this case, the property is quality and the indicator the number of citations.

2 Further references to the work of Robert Merton, especially his views on the notion of 'excellence', are found in Chapter 4 'Re-valuing research excellence: From excellentism to responsible assessment'.

3 Production and citation indicators can be used in a variety of ways that differ from the one pointed out here. I specifically refer to the use of these indicators for ranking purposes, a practice that has become popular in different research evaluation systems (Chavarro, Tang and Ràfols 2017).

4 This list is reproduced from Chavarro et al. 2017, and its sources are the reports SAB (2014) and SAB (2016).

5 These principles are not exhaustive and are given here only as an example to show some of the properties that are not addressed by dominant scientometrics evaluation.

6 https://sustainabledevelopment.un.org/sdg9

References

Chavarro D (2017) Universalism and particularism: Explaining the emergence and growth of regional journal indexing systems. Doctoral thesis, University of Sussex. http://sro.sussex.ac.uk/66409/

Chavarro D, Tang P and Ràfols I (2017) Why researchers publish in non-mainstream journals: Training, knowledge bridging, and gap filling. *Research Policy* 46(9): 1666–1680

Chavarro D, Vélez MI, Tovar G, Montenegro I, Hernández A and Olaya A (2017) *Los Objetivos de Desarrollo Sostenible en Colombia y el aporte de la ciencia, la tecnología y la innovación*. Documento de trabajo 01, Colciencias – UDEP http://www.colciencias.gov.co/sites/default/files/objetivos_de_desarrollo_sostenible_y_aporte_a_la_cti_v_3.5.pdf

Chavarro D, Ràfols I and Tang P (2018) To what extent is inclusion in the Web of Science an indicator of journal 'quality'? *Research Evaluation* 27(2): 106–118

Colciencias (2018) *Libro verde 2030. Política de ciencia e innovación para el desarrollo sostenible*. Bogotá: Colciencias. http://libroverde2030.gov.co/wp-content/uploads/2018/05/LibroVerde2030.pdf

Dollery BE and Worthington AC (1996) The evaluation of public policy: Normative economic theories of government failure. *Journal of Interdisciplinary Economics* 7(1): 27–39

Gallopín G (2001) *Science and Technology, Sustainability and Sustainable Development*. CEPAL. Retrieved from http://www.istas.ccoo.es/escorial04/material/dc12.pdf

Godin B (2006a) On the origins of bibliometrics. *Scientometrics* 68(1): 109–133. http://doi.org/10.1007/s11192-006-0086-0

Godin B (2006b) The linear model of innovation: The historical construction of an analytical framework. *Science, Technology, & Human Values* 31(6): 639–667. ftp://ftp.ige.unicamp.br/pub/CT010/aula%202/GODIN-The%20Linear%20Model%20of%20Innovation-STHV2006.pdf

Guba EG and Lincoln YS (1989) *Fourth Generation Evaluation*. Thousand Oaks, CA: Sage

Meadows DH, Meadows DL, Randers J and Behrens III WW (1972) *The Limits to Growth: A Report to the Club of Rome*. New York: Universe Books

Merton R (1973) The normative structure of science. In: N Storer (ed.) *The Sociology of Science*. Chigago and London: The University of Chicago Press. pp. 267–278

Molas-Gallart J and Ràfols I (2018) Why bibliometric indicators break down: Unstable parameters, incorrect models and irrelevant properties. *SSRN Electronic Journal*. January. https://papers.ssrn.com/sol3/papers.cfm?abstract_id=3174954

Mouton J (2014) Programme evaluation designs and methods. In: F Cloete, B Rabie and C de Conning (eds) *Evaluation Management in South Africa and Africa*. Stellenbosch: Sound Media. pp. 163–202

Nupia C (2018) La medición de la producción científica de los grupos de investigación en Colombia: del diálogo de expertos a la incorporación de prácticas más representativas. In: G Dutrénit and JM Natera (eds) *Procesos de diálogo para la formulación de políticas de CTI en América Latina y España*. Buenos Aires: Clacso https://www.researchgate.net/publication/325541711_La_medicion_de_la_produccion_cientifica_de_los_grupos_de_investigacion_en_Colombia_del_dialogo_de_expertos_a_la_incorporacion_de_practicas_mas_representativas

Ofir Z, Schwandt T, Duggan C and McLean R (2016) *Research Quality Plus (RQ+): A Holistic Approach to Evaluating Research*. Canada: IDRC. https://www.idrc.ca/sites/default/files/sp/Documents%20EN/Research-Quality-Plus-A-Holistic-Approach-to-Evaluating-Research.pdf

Orozco LA, Chavarro-Bohórquez DA, Olaya DL and Villaveces JL (2007) Methodology for measuring the socio-economic impacts of biotechnology: A case study of potatoes in Colombia. *Research Evaluation* 16(2): 107–122

Rijcke SD, Wouters PF, Rushforth AD, Franssen TP and Hammarfelt B (2016) Evaluation practices and effects of indicator use—a literature review. *Research Evaluation* 25(2): 161–169

SAB (2014) *The Crucial Role of Science for Sustainable Development and the Post-2015 Development Agenda. Preliminary Reflection and Comments*. UNESCO. http://en.unesco.org/un-sab/sites/unsab/files/Preliminary%20reflection%20by%20the%20UN%20SG%20SAB%20on%20the%20Crucial%20Role%20of%20Science%20for%20the%20Post2015%20Development%20Agenda%20-%20July%202014.pdf

SAB (2016) Science for sustainable development. Policy brief. UNESCO. http://unesdoc.unesco.org/images/0024/002461/246105E.pdf

Schot J and Steinmueller E (2016) *Framing Innovation Policy for Transformative Change: Innovation policy 3.0*. http://www.johanschot.com/wordpress/wp-content/uploads/2016/09/Framing-Innovation-Policy-for-Transformative-Change-Innovation-Policy-3.0-2016.pdf

UN (United Nations) (2015) *Transforming our World: The 2030 Agenda for Sustainable Development*. United Nations Organisation

Whitley R and Gläser J (eds) (2007) *The Changing Governance of the Sciences: The Advent of Research Evaluation Systems*. Dordrecht, the Netherlands: Springer

Wouters P (1999) The citation culture. Doctoral thesis, Universiteit van Amsterdam. https://pure.uva.nl/ws/files/3164315/8231_13.pdf

CHAPTER 13

Indicators for the assessment of excellence in developing countries

Rodolfo Barrere

Introduction

The accurate measurement of a certain phenomenon needs a concrete definition of its key characteristics and its boundaries. Measuring research excellence is therefore a major challenge because it can be defined in several ways, depending on the perspective and context. Generally, to be 'excellent' is to be superior in the achievement of a certain goal. In that sense, identifying excellence is to determine who has a better performance than others. The first step to tackle that challenge is to achieve a consensus on the goal. The second one is to find tangible expressions that can lead to its measurement. Another issue concerns the very concept of 'quality' related to excellence. The definition of quality, the criteria that express it and the indicators that would make it measurable are a theoretical problem, to which the solution is not simple. It is evident that there is no consensus about the content of the concepts of 'quality' or 'excellence' applied to research. How is quality translated into a variable that can be measured on a scale? (Albornoz and Osorio 2018).

In one way or another, scientific performance indicators are related to a concept of quality and can therefore be used to identify, categorise and 'measure' it. However, since quality is such an ambiguous concept, we usually work with indicators that describe the object of study without adjectives and, in this way, relate their characteristics – without ignoring their differences and particularities. In this sense, since indicators produce values or scores that can help quantify something that is difficult to measure, they contribute to the project of comparing diverse analysis objects, offering a 'translation' between a complex object and others, constructed in a theoretical framework in which its measurement produces a relevant meaning for the understanding of that object (Pérez Rasetti 2010).

The assessment of research excellence in low- and medium-income 'developing' countries has to be contextual. It can be seen in terms of quality, but also in terms of pertinence. For instance, while bibliometrics are a useful standard for knowledge production, they do not inform about other activities related to science and technology that can have a clearer impact on social needs. For example, scientific services (e.g. environmental monitoring, medical laboratory activities or engineering advisory) are not covered in commonly available indicators and so are not considered by policy-makers and funding agencies at the moment of evaluating groups or institutions. Also, the experience of a research group in knowledge transfer to social groups or to the business sector is commonly out of the scope. In order to move towards the proposal of a concrete set of tools, it is possible to define two separate fields where research excellence can be measured: one inside the scientific community and one outside the scientific community.

This chapter includes three main sections. The first section describes developments in Latin America to tackle specific characteristics of research and development (R&D) performance assessments. The second section discusses the use of traditional bibliometric indicators and bibliometric databases for the measurement of research excellence within the scientific community. Limitations of the most common international data sources are analysed and proposals for fostering journals in these countries are put forward. Finally, a set of indicators for the measurement of the engagement of researchers with society will be presented as an alternative for measuring research excellence outside the scientific community.

Background

Latin American countries show very different characteristics in terms of various items, ranging from their socio-economic indicators to the degree of consolidation of their science and technology (S&T) systems, as well as the maturity of their statistical systems. A wide gradient of situations exists within the region, including countries with features similar to those of the developed world and countries with very few R&D activities and an almost complete lack of statistical information. These diversities have been reflected within the sphere of the Ibero-American Network of Science and Technology Indicators (RICYT), which has worked as a discussion forum for S&T indicators since 1995.

Latin America is a heterogeneous region: two countries have a 'very high' score on the Human Development Index, while a third of the region is in the 'medium' group. The differences are also evident in R&D capacities. Only three countries (Brazil, Mexico and Argentina) are responsible for 92% of the regional R&D expenditure. Brazil expends 1.2% of its GDP on R&D, while many of the countries spend less than 0.15%. Some countries feature developed institutional systems and a complex set of policy instruments, while others have very incipient structures (RICYT 2017a). Science and technology systems in this context are also very heterogeneous, as are the demands from their societies. It is therefore a challenge to find a single definition of research excellence, as their goals and potential are very different. Governments are the main source of funds for R&D in developing countries, with the belief that it fosters social and economic development, but – even though we have the experience and methodologies to measure inputs and outputs of research activities – we still are unable to tackle the measurement of the social impact of science.

When RICYT was created, the availability of S&T information in Latin America revealed a problematic situation: most of the countries lacked reliable and comparable information. The initial feature of the network was to bring together two heterogeneous sets of actors: on the one hand, national S&T agencies, which are simultaneously producers and users of information and, on the other hand, researchers devoted to studying the relationships between science, technology and society, as well as experts in indicators. This duality conditioned both the focus and the agenda: it was a matter of generating indicators for policies and exploring new dimensions.

Producing indicators in Latin America is a task that involves not only transposing the methodological norms applied in developed countries, but also generating discussions in order to achieve consensus about which should be the more adequate indicators according to the intrinsic features of Latin American countries, without leaving aside international comparability. This involved two parallel tasks in the early years of RICYT. On the one hand, the OECD's methodological manuals were disseminated, with the aim of promoting international comparison. On the other hand, a discussion was generated around which necessary adjustments should be made to the manuals, in accordance with the idiosyncrasy of the region's countries. The debates referring to the more adequate methodological definitions for constructing input indicators, as well as the discussions on innovation studies, are clear examples of this situation. Nowadays, RICYT has developed a wide and active network that discusses methodologies and produces statistical information as inputs for decision-making and evaluation. That experience, in the diverse context of Latin American countries, is a good basis for the development of new tools for the assessment of research excellence in developing countries.

Excellence inside the scientific community: Bibliometrics

The use of quantitative indicators of research performance, especially those derived from bibliometric methodology, has become increasingly common for the evaluation of the scientific productivity of institutions and researchers, even in developing countries. The expansion of access and the facilitation of the use of these analytical tools and resources have generated a qualitative change in evaluation mechanisms. The possibility of, to a certain extent, automating evaluation through the use of bibliometric indicators is a temptation for those responsible for this activity, both because of its lower cost and easy management and to avoid overloading the researchers themselves.

Bibliometric indicators of knowledge production and utilisation processes – either research publications (publication output measures) or the citing of publications (citation impact measures) – are useful to measure the quality of research within the scientific community because the system of peer review (the assessment of colleagues themselves) guarantees its functioning. The scientific publication system, in addition to functioning as a reservoir of knowledge, is a prestigious distribution mechanism. In this sense, researchers seek to make their work known as widely as possible, using for that the most widely read (and cited) journals. The phrase 'publish or perish' is an adequate reflection of this phenomenon. In this context, prestige is an attribute that gets its meaning with regard to the work of colleagues; the peers in charge of the review will not recommend the publication of works that do not meet a minimum of quality and relevance.

This dual accountability mechanism ('publish or perish' and peer review) guarantees that the statistical analysis of scientific publications takes place in the context of the production of knowledge in an environment validated by the scientific community itself. The introduction of these assessment techniques, however, generates uncertainties about their influence on the behaviour of researchers (Hansson 2010), for example, on how researchers establish their research priorities and whether the choice of their line of work is conditioned more by the agenda of the high-impact factor of journals, rather than the relevance of the topic (at either the institutional or local level). In that sense, the most debatable issue is not the application of bibliometric techniques in developing countries, but the representativeness of the bibliographic databases on which those techniques are applied.

A common objection against the use of bibliometric indicators is related to a supposed weakness of international bibliographic databases with regard to their representation of scientific production in developing countries. The most common databases used in bibliometric analysis, such as the Web of Science (WOS) and SCOPUS, are multidisciplinary databases that are meant to be sufficiently representative of the mainstream of international science. The scientific, scholarly and technical journals indexed in those databases publish research on a range of subjects of interest at the international level and often include applications of common scientific techniques.

Nevertheless, a comparison between bibliometric indicators and statistical information generated by international organisations on the basis of national surveys of R&D activities shows a remarkable convergence. Sub-Saharan Africa (SSA) is responsible for 0.7% of global expenditure on R&D and has 1.1% of the researchers (see Figure 1). At the same time, 0.7% of the total articles indexed in SCOPUS are from SSA. In Latin America, the total expenditure on R&D represents 3.5% of global expenditure and the region has 3.9% of the world's researchers. Representation on SCOPUS is 4.5%. The comparison using WOS produces an equivalent outcome.

In this context, developing countries' contributions to mainstream science seem not to be under-represented. Nonetheless, the issues covered in indexed journals may not be the most important for developing countries. In that sense, there is a lack of robust bibliometric sources for a broader coverage of the scientific production of developing countries. There are no bibliographical bases capable of covering the entire scientific production of a country, which affects the possibility of using these sources for evaluation. This implies that the topics that interest the mainstream will be represented, while others will almost never appear. This phenomenon strongly affects developing countries, whose research topics, in some disciplines more than others, may diverge from those studied in leading countries.

The option of accessing regional bibliographic databases with a greater coverage of developing countries would allow a better representation of local research. Some Latin American initiatives aim to remedy this situation, such as the medical science database LILACS, developed by BIREME, and the CLASE and PERIODICA databases from Mexico's UNAM. SCIELO and REDALYC initiatives also offer encouraging prospects. However, there is still a long way to go. The statistical information available based on these regional initiatives still shows inconsistencies with the remaining available indicators, such as investment and human resources in R&D. Some countries are still over-represented, and others are under-represented in these regional data sources.

Figure 1: Percentage of global R&D expenditure, researchers and SCOPUS indexed articles

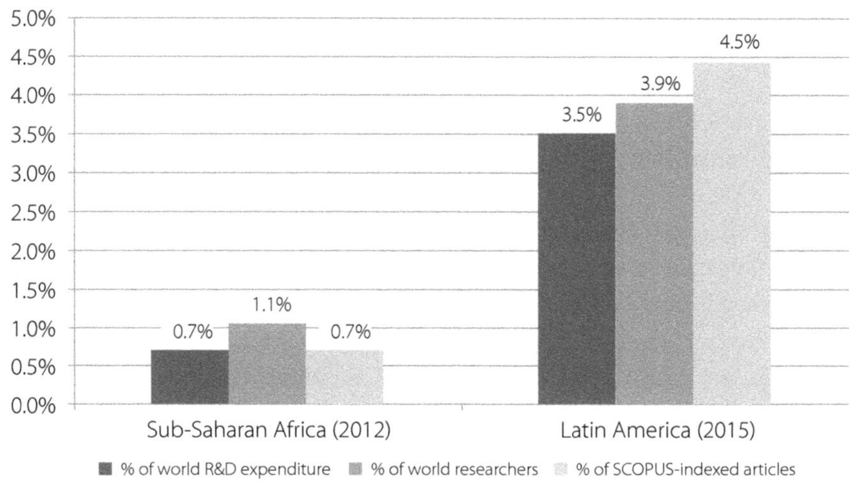

Sources: RICYT, UIS-UNESCO and World Bank (2014)

In conclusion, bibliometrics is a good methodology for measuring excellence within the scientific community, drawing on the need among researchers to publish and offering the quality assurance system through a strict peer review of submitted manuscripts. However, this assessment mechanism is only possible if journals meet the strict standards of editorial quality. In that sense, scientific journals which comply with editorial quality are valuable tools for the management and evaluation of S&T systems in developing countries. High-quality scientific journals help bring communities together and define agendas. However, most developing countries lack consolidated public policies for the support of fostering scientific journals. In Latin America, the few countries that have carried forward this type of policy, such as Brazil and Chile, are also the countries that have grown the most in their contribution to international science as measured in international bibliographic databases.

Beyond these general considerations, bibliometric indicators have broader limitations in measuring scientific production. Bibliometrics can only address the scientific aspect, while other activities and aspects, notably those of a technological, educational and social nature, must be studied by other indicators and information sources (Bordons 2001).

Excellence outside the scientific community: Engagement indicators

During the last decades there has been a growing demand from many governments – both in high-income 'developed' countries and other 'developing' countries – for academia to play a more active role in supporting economic growth and development. Universities, for example, are seen as key actors in their societies, because of their role in teaching, research and extension activities. These organisational missions have become part of the normative model of the 'modern university' in Latin American countries, but variations in the historical development of this model have produced different types of universities, each with their own specific profile, and operating in very diverse regional contexts.

Latin American public policies aimed at boosting economic growth, social development and increasing the efficiency of public management have placed the focus on innovation. This is underpinned by the understanding that innovation is the result of a synergistic engagement and action involving several organisational actors – including universities and other public research centres – to transfer knowledge, skills and other capacities to society. Universities are seen as key players in innovation systems.

The experience of RICYT with its Bogotá Manual, which is focused on innovation, shows that a typology of Latin American firms is different from that of European firms and the industrialised world in general. Likewise, available indicators highlight that the role of universities in the production of knowledge is central in Latin American countries, in comparison with other regions, in which the impulse of the business sector predominates. For example, in Latin America, 75% of the total researchers are based in universities, compared to only 39% in the European Union. Regarding universities' share in

SCOPUS-indexed research articles, in Brazil, Chile and Colombia, for instance, this is near 90%, while in European countries it is usually less than 70% (OEI 2018).

The high percentages of poverty in Latin American countries also present a picture of social demands; this challenges academia in a different way than in countries with a higher degree of development. In this context, many Latin American countries and governments have implemented policies to encourage collaboration between academia and the business sector, as well as initiatives to finance scientific infrastructures, with the purpose of contributing to the transfer of research results to the whole society. To monitor and manage this process, there is a need to design, develop and implement a system of indicators capable of reflecting a wide range of interactions through which academia relate to their socio-economic environment. Following this requirement, RICYT sought to provide an answer. From its beginnings, in 1995, RICYT had in the foreground the challenge of measuring the social impact of science and technology. In these discussions, the link between academia and the socio-economic environment has repeatedly appeared as one of the mechanisms through which this impact is made effective. The *Ibero-American Manual of Engagement Indicators of the University with the Socioeconomic Environment* – the Valencia Manual (RICYT 2017b) arose as a result of a long process of reflection that sought to respond to a demand for accurate and comparable information regarding the influence of the universities on the socio-economic environment. The initiative was driven by the Ibero-American Observatory of Science, Technology and Society (OCTS) of the Organization of Ibero-American States (OEI) and RICYT, with the support of Centro REDES in Argentina and INGENIO (CSIC-UPV) in Spain.

Opting for the university as an observation point and unit of analysis is related to the above-mentioned role of these institutions in the various research systems in Latin American countries. The proposal also includes the possibility of observing engagement patterns at the level of the academic groups at the base of the university organisational pyramid; that is, the possibility of analysing the behaviours of academics in terms of their links with external actors and detecting non-institutionalised linkages.

To define the scope of the Manual, 'engagement activities' are understood to be those related to the following:

- The generation of knowledge and the development of capacities in collaboration with non-academic agents and the elaboration of legal and cultural frameworks that guide the opening of universities towards their environment; and
- The use, application and exploitation of knowledge and other capacities existing in the university outside the academic environment, as well as training, sales of services, advice and consultancy, carried out by the universities in their environment.

The indicators proposed in the manual are, in general, quantitative measures, although in some cases qualitative descriptions are used to facilitate the interpretation of the development of the engagement activities within the environment of each institution.

The set of proposed indicators is grouped into three categories:

- *Institutional characterisation:* these indicators refer to aspects indirectly related to the engagement activities that facilitate and condition their existence and development in the institution (such as the history of the institution, its size and its profile of academic specialisation), which are relevant in characterising the institutional context and appropriately contextualising the activities of engagement;
- *Indicators based on the capacities for the engagement activities:* the engagement activities of each institution are based to a large extent on the use of the available capacities. These indicators account for the stock of knowledge, as well as the capacities associated with the physical and organisational infrastructure of each institution. Some examples are intellectual property rights, infrastructure marketing and spin-offs and start-up creations; and
- *Indicators based on the engagement activities themselves:* although knowledge of the characteristics of the institutional organisation

and of the available capacities is central to understanding the link between the university and the environment, the intensity with which these activities take place in the institution is observed directly in the range of engagement activities carried out. This group of indicators is meant to capture the effective realisation of these activities, and the results obtained from them. Examples include the number of contracts in collaboration with different sectors, capacity-building activities developed, extension activities and the social communication of knowledge.

In principle, information that leads to a global characterisation of the institution is requested. This includes the interaction with the environment carried out by its different academic units, which reveals institutional patterns in terms of the type of activity, financing methods, resources generated and socio-economic sectors with which it is linked. Having specific information and dedicated indicators on such university–society interactions is of fundamental importance, on the one hand, in order to provide academic institutions with instruments to measure their own engagement activities and, on the other hand, to provide governments with instruments that allow them to design public policies and define the strategic allocation of associated resources that accompany them. Also important is the use of information by different economic and social actors to guide their strategies for finding links with universities and academic groups. It is also necessary that such indicator systems take into account the specificity of the social and productive landscape of developing countries and the characteristics of their universities and public research centres. The decentralised nature of university engagement activities within the socio-economic environment poses a significant challenge to the collection of information. The need to have an adequate information system on these activities is thus a fundamental step for the development of a system of indicators that is broad enough to cover the greatest number of aspects related to the link between the university and the environment in the specific context of each institution.

A pilot study was carried out in six universities in five Latin American countries. Although it was exploratory work which had the objective of perfecting the methodology, the results offer some interesting clues about the links in the universities of the local region, which should be deepened in later studies (Estébanez 2016). Findings indicate that both the execution and the management of the engagement activities take place in multiple institutional spaces within these universities. Each case shows different patterns in terms of the efforts made in the engagement activities, with varying degrees of importance in relation to other activities, such as R&D.

The most standardised management modality of engagement activities is the contract. In this regard, very diverse activities are carried out, some involving the generation of new knowledge and others that are routine services. There are contracts for research, training of human resources, technological development and technology licensing.

In addition to producing a preliminary diagnosis of engagement activities in regional universities, the application of the pilot study yielded a series of conclusions regarding the methodological strategies to be implemented in future surveys, and associated possibilities and limitations in data collection. The development of engagement indicators will be of great interest to better understanding relationships between universities and wider society. One of the main methodological and analytical challenges was the difficulty in capturing linkages at the research group level; such activities are usually very rich, but often not registered at higher levels within the university. Next year's OCTS is planning to apply a massive regional online survey that targets academic authors in order to gather this 'micro-level' information in a comparative way.

General conclusions

In the context of S&T systems management, and for the allocation of resources, research excellence cannot be defined in a single way. This complex concept depends on desired results and impacts. However, it is possible to define different domains of application where excellence can be defined and measured, each domain with its own logic and quantitative aspects. As was discussed previously, one possibility is to separate excellence measurements inside and outside the scientific community.

Looking at research excellence from the perspective of the scientific community, bibliometric indicators have proved to be valuable analytical tools with consolidated methodologies that nowadays have permeated research activity itself. The available information sources and international databases are sufficiently accurate to measure the contribution of developing countries to mainstream science, but additional databases are needed, or need to be developed, for a broader measurement of knowledge production that also captures local and regional dimensions. To make that possible, it is also necessary to develop a strong scientific journal system, which includes more local and regional journals, compliant with high-quality editorial standards. This is a public policy vacancy in most developing countries.

Measures of research excellence should also include university–society engagements; this is where significant impacts of R&D investments are to be expected. That is an important challenge, as links have very different forms and are not always recognised in the institutions. The Valencia Manual methodology is an interesting collective experience to tackle that challenge.

These different dimensions of measuring excellence, and many others that can be defined, are by no means mutually exclusive. They offer complementary approaches that provide a broader landscape in which the results and evaluation of research activities can be viewed. The ideal research project is one that can show excellence in many dimensions, depending on the goals established by funders, donors or policy-makers.

References

Albornoz M and Osorio L (2018) Rankings de universidades: calidad global y contextos locales. *Revista CTS* 37(13). http://www.revistacts.net/volumen-13-numero-37

Bordons M (2001) Aspectos metodológicos en la obtención de indicadores bibliométricos. *Cuadernos de Indicios* 1

Estébanez M (2016) Medición de las actividades de vinculación de las universidades con el entorno. Aplicación piloto del Manual de Valencia. Lugar: Buenos Aires. http://www.ricyt.org/files/Estado%20de%20la%20Ciencia%202016/

E2016_2_3__MEDICIN_DE_LAS_ACTIVIDADES_DE_VINCULACIN_ENTRE_LAS_UNIVERSIDADES_Y_SU_ENTORNO__UN_ANLISIS_REGIONAL.pdf

Hansson F (2010) Dialogue in or with the peer review? Evaluating research organizations in order to promote organizational learning. *Science and Public Policy* 37(4)

OEI (Organization of Ibero-American States) (2018) Las universidades, pilares de la ciencia y la tecnología en Améria Latina. http://observatoriocts.org/files/Archivo%20Documental/Libros%20del%20Observatorio/CRES2018.pdf

Pérez Rasetti C (2010) Construcción de indicadores para el sistema de educación superior de iberoamérica/américa latina y el caribe. Reflexiones para una propuesta. Observatorio Ciencia, Tecnología y Sociedad (OCTS/OEI)

RICYT (2017a) El Estado de la Ciencia en Iberoamérica. RICYT. http://www.ricyt.org/publicaciones

RICYT (2017b) Manual Iberoamericano de Indicadores de Vinculación de la Universidad con el Entorno Socioeconómico - Manual de Valencia. http://www.ricyt.org/manuales/doc_download/152-manual-iberoamericano-de-indicadores-de-vinculacion-de-la-universidad-con-el-entorno-socioeconomico-manual-de-valencia

World Bank (2014) *A Decade of Development in Sub-Saharan African Science, Technology, Engineering & Mathematics Research*. Washington DC: World Bank. http://documents.worldbank.org/curated/en/237371468204551128/pdf/910160WP0P126900disclose09026020140.pdf

Statistical sources

Ibero-American Network on Science and Technology Indicators (RICYT): http://www.ricyt.org
UNESCO Institute for Statistics (UIS-UNESCO): http://uis.unesco.org/

CHAPTER
14

Rethinking scholarly publishing: How new models can facilitate transparency, equity, efficiency and the impact of science

Liz Allen and Elizabeth Marincola

Introduction: Legacy publishing systems and requirements for new approaches

For centuries, the scientific journal has been the medium through which original research findings are reported and disseminated. Up to the late twentieth century, space and cost restrictions, dictated by printed copy formats, were main drivers for scientific publishers to develop processes to help them decide and prioritise what to include in a specific journal volume. However, over time, the development of selection criteria and processes used to identify content that publishers would wish to include in their journals has morphed considerably and is now thought to have had a detrimental effect on the careers of scientists and on the progress of science more broadly. And for researchers working in the Global South and in resource-poor environments, the detrimental effects are thought to have been particularly acute, presenting significant barriers to entry to publish in a highly selective journal market.

Today, scientists from across the world experience significant frustration with both the requirements and processes involved in sharing and disseminating the results of their research. Studies show that many of the processes and practices used by legacy publishers, which may have been somewhat justifiable in the era of printed journals, are outdated and outmoded, dependent upon complex, cost-inefficient, opaque, time-consuming processes that are largely non-transparent and, taken together, are a significant cause of research funding waste (Chalmers and Glasziou 2009; Chan et al. 2014; Munafò et al. 2017). Delays of months or more between submitting an article and it being published, access and licence constraints, bias resulting from opaque peer review, the tendency to favour the publication of positive results and incomplete availability of data are among the myriad issues that face a researcher wanting to publish research (Warren 2003; Harris et al. 2006; Carrol et al. 2017). Moreover, the determination to publish is often driven by subjective criteria about whether an article contains novel, exciting or radically new perspectives – what has implicitly come in selective journals to define 'excellence'. Such selection criteria result in the fact that much important, useful work, performed at public expense, is being left unpublished (including for example, negative and null findings). In addition to the resultant waste of resources and the hindering of careers, the advancement of science itself is in large part dependent on the building of such 'incremental' results.

Furthermore, the promise and potential for cost reductions in scholarly publishing associated with a shift to largely digital formats (i.e. not print) do not seem to have been passed on to the researcher and consumers of scholarly output, evidenced by escalating costs for publication and journal subscription fees.[1] Added to this, information about what (and predominantly *where*) someone has published remains the dominant currency used across the world to support research and researcher evaluation, informing grant allocation, and career appointments and promotions for researchers and research teams. The need for researchers '*to publish and to publish well*' therefore creates a reliance on an established publishing system and an inertia among publishers to change the service or the status quo.[2]

In today's world of the web, the costs of space – for paper, printing, shipping and storage – once incurred by scholarly publishers have largely disappeared. Moreover, the 'costs' to the reader of combing through detailed information within an article and the magnitude of articles published in one's field have also largely disappeared, thanks to powerful search tools that enable consumers to zero in efficiently on content of interest. These forces demand that it is no longer acceptable to limit the sharing of science output through selection which is too often subjective and arbitrary. It is time to reinvent outmoded and potentially damaging publishing practices and policies. This particularly applies to the selection process of original research for publication, encapsulated in the recent statement from the leadership of the Howard Hughes Medical Institute (HHMI) that scholarly publishing should move to a system of *'publish first, curate second'* (Stern and O'Shea 2019). The frontier of science communication must be an approach that *combines* the ability of researchers to publish rapidly, without pre-selection according to interest and novelty, with a mechanism to assure quality and trust in the work being published through peer review that is open and transparent throughout. The overall goal is to accelerate access to original research findings of all types, in order to optimise the use, re-use and potential impact of research – indeed to incentivise its creation in the first place. Enshrined in such an approach is the belief that researchers (as authors, users and consumers of research) are in control, thus removing the barriers to publish that disproportionately affect researchers from less-established research institutions or resource-poor environments.

The changing landscape of scientific publishing

Prior to the introduction of the internet, dissemination of knowledge was relatively slow, dictated by largely manual processes for selecting, validating, editing, setting, printing, mailing, archiving and storing research journals. As the system for cataloguing and recording the use and citation of published material developed, so the practice of developing and using bibliometric indicators around how published research was 'used' (i.e. cited) by others became a key component of how a researcher's productivity and 'excellence' was judged. Today, the Journal Impact Factor (JIF) of the journal in which a piece of work is published remains a remarkably sticky (despite being generally agreed as being misapplied and unhelpful [Zhang et al. 2017]) proxy for the quality of published work.

Since the late twentieth century, enabled by the web, the volume and diversity of research output being published began to increase rapidly, and continues to grow. Access to research has been further facilitated by the introduction of open access (OA) business models across scholarly publishing, first introduced by BioMedCentral (BMC) in 1998,[3] supported by requirements and mandates from research funding agencies and research institutions for scientists to make their work available in OA formats. There have been a variety of responses over time from states and regions across the world to support OA for publicly funded research findings; see, for example, the SciELO[4] and Redalyc[5] initiatives in Latin and South America and the recent cOAlition S 'Plan S'[6] in Europe. Market share of OA in STM publishing has grown since its introduction to about 12% of articles and 26–29% of journals as of 2017.[7] However, despite many funding agency and institutional requirements and policies to encourage and mandate researchers to share their research through open access, achieving OA as a global standard remains elusive for many practical, economic, cultural and political reasons, compounded by a system of scholarly publishing which has been slow to adapt to the requirements of a digital, OA world.

Despite the growth in the capacity to publish, most publishers continue to hold on tight to their role of custodian and gatekeeper of what science is eventually published in their journal, in large part because, whether they are commercial or non-profit publishers, they must be at least financially sustainable, and in many cases profitable. In the case of non-profit publishers, such as scientific societies, journal revenue often sustains the other activities of the organisation. Over time, many publishers have grown into large corporate enterprises that are accountable to stakeholders and driven by a profit motive that means that the interests of the entity and its stakeholders are often at odds with the interests of scientists and the advancement of science more broadly. The dominant role of scholarly publishers on science communication practices has been hard to loosen, in great part because hiring, grant-making, promotions and awards have been determined by *where* a researcher has published instead of *what* is being described and the intrinsic value of the insights being published. The fact that many large scholarly publishers are governed by the vested interests of their company shareholders and by profit margins makes for a system that is unlikely to have the interests of science and scientists as its first priority. While individual scientists usually recognise the dysfunction of this system, they generally feel that they are hostage to it, especially early-stage researchers who are dependent on the system to gain a foothold in their career.

For all these reasons, publishing practices are replete with outdated and unfair features. First of all, over time, the judgement of a small number of editors as to the ground-breaking nature, novelty and 'excellence' of research – as indicated by its selection for publication – has proved weak at best. This is not because of any lack of intelligence of editors, but rather because the nature of research is such that it is difficult, if not impossible, in most cases to determine *a priori* what the value of a particular research output will be after it is (or is not) built upon by others. When making a 'value' calculation, it is moreover important to bear in mind that the *ultimate* value of research is its return to the taxpayer (who is the major funder of research), other funding agencies that invest in research, and of course individuals whose well-being depends on it – as measured in human health, agricultural and veterinary advances and environmental benefits. Second, the traditional curatorial function of editors – to comb through many submissions to select the nuggets that they think will be of greatest interest to the greatest number of scientists who may read it – is much less essential now that search tools can in seconds enable scientists to home in on findings of specific interest better than any editor or group of editors possibly could. And third, a huge amount of scholarly output is wasted: because it ages beyond a useful point while awaiting journal acceptance, because most of it is still hidden by subscription barriers from most other researchers who are thus prevented from building upon it, because of limitations in the form and nature of publishable outputs, because of peer review that is only accessible to authors and because of the failure to require that the data upon which claims are staked be shared with others who may wish to analyse, collaborate and/or reproduce findings.

Nevertheless, driven by a number of influential research funders, institutions and research leaders, change is coming; change that is likely to significantly reduce the barriers to entry to share and publish scholarly work. And change that is significantly likely to benefit those who to date have found it difficult to compete and have equitable access to a scholarly publishing system founded upon criteria of being highly selective and driven by subjective notions of 'excellence' and novelty.

Science communication at an evolutionary inflection point

It is evident to those in the field of science publishing and to many scientists worldwide that for these myriad reasons, traditional scientific journals themselves are an outdated mode of building research to the benefit of humankind. Yet, as is predictable with a product that has been the standard – indeed to many people, the only imaginable mechanism for stimulating, rewarding and building science – for over three centuries, it is hard to abandon, notwithstanding widely recognised shortcomings. First, loyalty to the concept of traditional publishing, as well as to particular journals, is extremely strong. Declaring that an author has had a 'paper' published in a highly selective journal is in itself often used as shorthand for success and prestige. The prestige of any particular journal has come to be measured by the handy yet misleading JIF. It is very hard to compete with the brand value that the highest JIF journals offer, especially in the crowded marketplace of scientific output. Second, editorial boards, as well as staff editors, identify strongly with the title(s) with which they are associated, often especially so when the titles are published by the disciplinary scientific society to which they have a parallel loyalty. And third, anyone who has ever published in a particular journal during the course of its existence has a vested interest, as well as often an emotional bond, to the journal that conferred prestige on the author by accepting his or her paper for publication.

Yet journals, while they have enjoyed an impressive run, are no longer necessary – at least not in their current format as the dissemination point for original research. Indeed, other mechanisms are potentially much more effective vehicles for the sharing, discovery and dissemination of research results. We have seen the introduction and massive growth of the 'mega-journals' (spring-boarded with the introduction of *PLOS One* in 2006),[8] which are designed to select and publish content based only upon 'soundness', reducing the editorial function of the journal to a focus on credibility of the work. More efficient tools and services now exist to support discoverability of content and articles; journal editions often contain such a mix of articles that readers are unlikely to be interested in the full range of articles in one edition, and are much more likely to search for specific articles or material to use through bibliographic and citation databases such as PubMed (for biomedical research) and/or Google Scholar, Scopus or Web of Science.

It is incumbent upon policy-makers, governments, foundations, universities, science disseminators and public-interest entities to fully displace any use of the JIF with a more rounded and tailored suite of research-related indicators that can be used to support decision-making, in all its guises, across the industry of sciences, as advocated through DORA.[9] This will clear the way for researchers to publish, share and collaborate around scientific findings in a manner that will speed up the progress of science and increase the fairness of the system used to judge researchers for grants, awards, tenure and promotion. This in turn will enable funders to maximise the value of their research investment.

A solution? The growth of rapid and open publishing platforms

A number of new approaches to scholarly publishing have emerged in the last years, particularly those focusing on the demand among research stakeholders to make findings more accessible at speed. Perhaps the most notable growth has been in the use by researchers of rapid publication platforms such as those provided by pre-print servers such as arXiv, bioRxiv and the open research platforms provided predominantly by F1000 (see Figure 1 which presents the lens of bioscience-related content).

Figure 1: Growth in content being published via rapid publication models

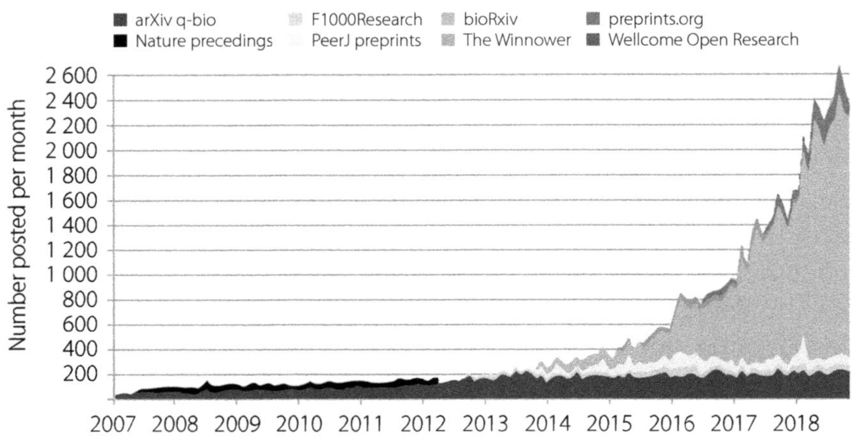

Source: http://www.prepubmed.org/monthly_stats/

In 2013, F1000 introduced the first open publishing platform for science, F1000Research, which effectively *combined* the key benefits of preprinting (*rapid publication)* with peer review (*quality assurance* provided by experts). The approach removes any undue delay with the publication of submitted research, publishing after a series of basic technical, ethical and credential checks (which remains an important component of the validation of research), but before invited peer review is undertaken, termed *post-publication peer review.* F1000 has since worked with a range of funding agencies and research-performing institutions, as its publishing approach chimes with their increasing demands for more rapid and open access scholarly publishing services that present minimal barriers for those wishing to publish, while being cost effective.

And outside the Global North and high-income countries (HICs), and perhaps where research assessment systems are less entwined with a focus on scholarly publications, the introduction of new outlets for sharing research presents an opportunity for scientists. In Latin America, facilitated by the SciELO network, open access to research produced in the region is simply not an issue, as all content is guaranteed OA. Publication of research on preprint servers across the world provides an easy route for researchers to present early sight of their work. Initiatives to secure full OA, such as Plan S, led by the cOAlition S funders across Europe, are driving the reinvention of legacy publishing systems; this reinvention has resulted in newer models of publishing becoming viewed as part of the mainstream. Researchers in the Global South are effectively in a position to 'leapfrog' over much of the legacy system and take advantage of and help shape a new world of scholarly publishing.

There are examples in other sectors that demonstrate how later adopters of technology can leap ahead, bypassing legacy systems and processes, notably in banking, fintech and utilities. In Kenya, more than 90% of the population were without a bank account at a time when 88% of individuals enjoyed access to a mobile phone. Thus Mpesa was introduced, with little notice or resistance from the banking industry, enabling people to move money through mobile platforms. Today, 60% of Kenyans actively move money – from buying bananas from a street vendor to paying for a vehicle – through Mpesa.[10] However, there is only isolated uptake of such technology in the US and Europe.

In recent years, while the quantity and quality of research output has rapidly grown across Africa, the role of legacy of science publishers within this has not been dominant. African scientists have traditionally found it difficult to publish work in journals based in the Global North because of the lack of familiarity among editors with regard to African laboratories and institutions, and the perception of the lesser value of more locally based research findings to a global audience. In 2018, recognising the opportunity to bypass legacy publishing systems for scientists in Africa and to help build reach to findings and research capacity, the African Academy of Sciences took the bold step to launch AAS Open Research. AAS Open Research joined a group of other funder-sponsored open research platforms to demonstrate that new models of publishing (outside of traditional journals) can help to deliver good science that is fully accessible and useable by all.

How open research platforms work: Case study of African Academy of Sciences Open Research

The guiding principle of open research publishing platforms such as

AAS Open Research, provided as a service by F1000, is that they are *'author-centric'*: shifting the balance of power about what is published to the authors and away from publishers. Figure 2 presents an overview of the publishing process adopted by AAS Open Research.

Importantly, the approach, like a mega-journal, is content agnostic and takes a holistic view of the types of research output it can publish – there are no space limitations and there are no editors screening out content for interest. Published outputs include not just traditional research articles, but any research output that requires peer review, including methods, study protocols, software tools, case reports and research notes. While fully open, peer review is not 'crowd-sourced' but is invited, and a published piece is considered iterative, not static: versioning is clearly delineated and each revision is entirely open and visible throughout, moving to a concept of continuous publishing.

Figure 2: Overview of publishing process used by AAS Open Research

Immediate & transparent publishing

AAS Open Research provides researchers supported by AAS and programs supported through its funding platform, AESA with a place to rapidly publish any results they think are worth sharing. All articles benefit from immediate publication, transparent refereeing and the inclusion of all source data.

Our publishing process

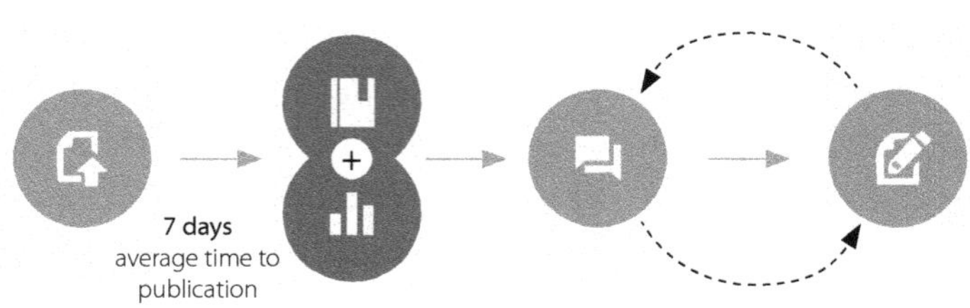

Article submission
Submitting an article is easy with our single-page submission system. The in-house editorial team carries out a basic check on each submission to ensure that all policies are adhered to.

Publication & data deposition
Once the authors have finalised the manuscript, the article (with its associated source data) is published within a week, enabling immediate viewing and citation.

Open peer review & user commenting
Expert referees are selected and invited, and their reports and names are published alongside the article, together with the authors' responses and comments from registered users.

Article revision
Authors are encouraged to publish revised versions of their article. All versions of an article are linked and independently citable.

Source: https://aasopenresearch.org/about

This transparency applies to the content as well as the reviews, offers credit and exposure not just to authors, but also to reviewers; it also invites the active or passive participation of readers who can benefit from the content of scientific exchange between authors and reviewers. Published work is subject to quality control through the invited peer-review process and is fully indexed after it passes peer review. However, even before indexing, the work is available for others to see and scrutinise (as it is on a pre-print server).

As part of a move to a world of 'open publishing', F1000 has been working alongside its platform partner to help shape a paradigm for how original research should be published in the future to maximise its potential for use and to minimise the risk of waste, duplication and redundancy. We consider there to be a number of key requirements for work being published via such an 'open research' mode of publishing in order to help to assure its provenance, credibility and trust, and thereby its rigour and potential for use and re-use (see Table 1), though these remain a work in progress. Some of these features (e.g. open access, FAIR data) are finding their way into legacy publishing systems. All of these – and more? – could be essential to underpin a more transparent, equitable, efficient and impactful science publishing system for the future, and one that removes the barriers to publication of research for researchers in the Global South. And we are keen to build the evidence base around how new models work best to support researchers across disciplines, career levels and geographies.

Indicators of quality and importance

In all the opportunities presented by new modes of rapid and open publishing, it is important to remember that researchers still require indications of their productivity and quality of research output and impact. Research outputs, in all their forms, remain a valuable contribution to knowledge, and are the route through which researchers share and communicate their progress and discovery. Such indicators are also vital for users of research findings, such as health professionals, journalists and policy-makers, to help get relevant research findings into policy and practice more effectively and without unnecessary delay. It

Table 1: Principles for the publication of original research in an 'open' future

Principle	What?	Core aim
Pre-review publication	• all submissions published prior to peer assessment for research quality	**speed of access to new knowledge**
No selection	• all submissions assessed against only objective technical checks, e.g. plagiarism, ethics, readability, scope • no subjective checks for novelty, perceived importance or impact	**reducing reporting bias and publication bias**
FAIR source data and resources	• underlying source data/software made FAIR (findable, accessible, interoperable and re-usable) • adhering to the principle of 'as open as possible, as closed as necessary'	**maximising re-use and enabling verification**
Open access	• immediately OA: open, machine-readable licensing that enables re-use for any purpose, subject to attribution	**maximising access and potential for use and re-use**
Open, signed, invited peer review	• peer review by invited experts; conflicts of interests declared • peer review reports openly published, and reviewers named, with the ability to publish new versions • peer reviewers able to get visibility and credit for their efforts in supporting the work of others	**transparency, fairness and accountability**

remains important that there be credible measures of the value, importance, use and re-use of research findings and data.

Research outputs published outside the traditional journal system, but which secure a digital footprint (e.g. digital object identifiers [DOIs]) and bibliographic record – such as are made available through pre-print servers and open publishing platforms (e.g. *AAS Open Research*) – are as discoverable, trackable, citable and useable as those published within the traditional journal system, except they can be reached and discovered more quickly and openly.

Furthermore, open-peer review is increasingly helping to support visibility and recognition of the work that scientists do as 'peer reviewers' in supporting the development of work being published through initiatives such as ORCID and Publons. And, in actual fact, transparent refereeing provides researchers and potential users of research with another marker of quality as a peer reviewer's credentials; what they say about a piece of research can become part of the assessment, as well as part of the reviewer's own scientific output, instead of hidden and lost from the public record.

Conclusion

There is massive change afoot in the scholarly publishing and communication system. The balance of power is shifting as researchers, funders and institutions are demoing more rapid access and usability of research findings. We know that many of the processes and systems intrinsic to traditional science publishing are increasingly outdated and anachronistic. And we know that the ecosystem for research and researcher evaluation and assessment has been built upon an unhealthy and misleading dependence on indicators of research quality and value, based largely upon a judgement about *where* someone has published their work instead of *what* has been discovered and how research might have value in all its forms. But we believe that this is changing. New modes and outlets for sharing research outputs are reducing the practical barriers for researchers from across the globe wishing to share their findings and participate in a more connected and open science system.

In the absence of complex research assessment systems focused upon scholarly publications and many of the constraints and legacy systems and processes that researchers in HICs face, researchers in the Global South are in a good position to 'leapfrog' over established systems of publishing and to take advantage and help shape this new world of scholarly publishing. Adoption of models with features such as those integral to AAS Open Research can put researchers in control of what they wish to share, and enable the publication of insights and findings that are important in a global, regional and local context, no matter how ground-breaking or 'excellent'. And, most practically, changing the paradigms and basis upon which research is *selected* for publication effectively frees up researchers to be honest and holistic in what they share.

Many of the challenges in making this shift are more philosophical, financial and political than technological. They involve rethinking how stakeholders can work together to provide solutions and services that can be best tailored to support rapid, shareable publication and access to research findings. We believe that this challenge signals an inflection point that presents researchers in low- and middle-income countries – where dependence on legacy publishing systems, cultures and assumptions is less of a barrier than elsewhere – with unique and important opportunities to lead the way.

Research has value in many different ways and in many different contexts; that why it is done in the first place. Communicating what is found (or not) during research, and most especially when this involves the use of scarce resources, is a core requirement of the research process. It is, and always has been, an essential part of the research process; remodelling how that research is shared and published to improve access, to enable use and re-use and to reduce waste makes for an effective and efficient science system – with benefits for all concerned.

Notes

1 See for example: https://wellcome.ac.uk/funding/wellcome-and-coaf-open-access-spend-201617
2 See analysis: https://www.forbes.com/sites/stevensalzberg/2019/04/01/nejm-says-open-access-publishing-has-failed-right/#31b8b0d76a44
3 https://blogs.biomedcentral.com/bmcblog/2015/10/22/history-open-access/
4 http://www.scielo.org/
5 https://www.redalyc.org/home.oa
6 https://www.coalition-s.org/
7 https://www.zbw-mediatalk.eu/wp-content/uploads/2017/07/STM-Report.pdf
8 https://www.plos.org/history
9 https://sfdora.org/
10 See: https://lup.lub.lu.se/student-papers/search/publication/8952015

References

Carroll HA, Toumpakari Z, Johnson L and Betts JA (2017) The perceived feasibility of methods to reduce publication bias. *PLoS ONE* 12(10): e0186472

Chalmers I & Glasziou P (2009) Avoidable waste in the production and reporting of research evidence. The *Lancet* 374: 86–89

Chan AW, Song F, Vickers A, Jefferson T, Dickersin K, Gøtzsche PC et al. (2014) Increasing value and reducing waste: Addressing inaccessible research. The *Lancet* 383: 257–266

Harris IA, Mourad MS, Kadir A, Solomon MJ and Young JM (2006) Publication bias in papers presented to the Australian Orthopaedic Association Annual Scientific Meeting. *ANZ J Surg.* 76(6): 427–431

Munafò MR, Nosek BA, Bishop DVM, Button KS, Chambers CD, du Sert NP et al. (2017) A manifesto for reproducible science. *Nature Human Behaviour* 1. https://www.nature.com/articles/s41562-016-0021

Stern BM and O'Shea EK (2019) A proposal for the future of scientific publishing in the life sciences. *PLoS Biol* 17(2): e3000116. https://doi.org/10.1371/journal.pbio.3000116

Warren B (2003) Current challenges and choices in scientific publication. *Proceedings (Baylor University. Medical Center)* 16(4): 401–404

Zhang L, Rousseau R and Sivertsen G (2017) Science deserves to be judged by its contents, not by its wrapping: Revisiting Seglen's work on journal impact and research evaluation. *PloS One* 12(3): e0174205

CHAPTER 15

Research Quality Plus: Another way is possible

Jean Lebel and Robert McLean

Why Research Quality Plus?

In India, the world's leading producer of mangoes, up to 40% of the harvested fruit is destroyed in transit before delivery. This costs up to USD 1 billion in lost income each year, affecting the lives and livelihoods of millions of farmers, traders and consumers. So researchers from India, Sri Lanka and Canada developed a suite of nanomaterials that can be sprayed onto fruit on the tree, in packaging or in transit, to extend its life. They trapped hydrophobic hexanal molecules (derived from plant waste) in a hydrophilic membrane so that they could be suspended in liquid for application to the fragile fruit.

In Egypt, more than 95% of women have experienced sexual harassment at least once, and most cases go unreported. In 2010, researchers at the Youth and Development Consultancy Institute in Cairo developed Harrassmap. This online interactive resource enables people to report and map cases of sexual harassment. When it emerged that university campuses were hotspots, Cairo University implemented a policy to combat sexual harassment, the first of its kind in the Middle East. Other universities in Egypt are following suit.

Both projects help to solve pressing societal challenges. The researchers involved appreciate that the people who benefit from the projects are the ones who are best placed to judge the value and validity of the work. The research teams spent time developing their hypotheses and results with those who feel the effects. In each case, the research is robust and life-changing – exactly the combination that most people would say is the very purpose of science.

But both projects would score poorly if judged using only conventional approaches to evaluating research quality that prioritise the opinion of peers, the volume of papers published and citations. That's a problem because it is endorsement from other scientists, not stakeholders, that drives career advancement for researchers in Egypt, Sri Lanka and India, as everywhere else.

Is the weakness in the science or in the way it is measured? Too often it is the latter, in our view. Dominant techniques of research evaluation take a narrow view of what constitutes quality, thus undervaluing unique solutions to unique problems. At Canada's International Development Research Centre (IDRC) in Ottawa, we fund just this sort of research: natural and social science that unearths fixes for the development challenges facing countries in the Global South. The majority of the work we support is led by researchers from these countries.

So we at the IDRC developed a tool to evaluate the quality of research that is grounded in, and applicable to, the local experience. We used it to assess 170 studies and then did a meta-analysis of our evaluations. The results suggest that it is possible – and essential – to change how we assess applied and translational research. We call it the Research Quality Plus, or RQ+, approach.

Tunnel vision

The limitations of dominant research-evaluation approaches are well known (ASCB 2012; CIHR 2013; Hicks et al. 2015; Wilsdon et al. 2015; Holmes 2016). Peer review is by definition an opinion. Ways of measuring citations – both scholarly and social – tell us about the popularity of published research. They don't speak directly to its rigour, originality or usefulness. Such metrics tell us little or nothing about how to improve science and its stewardship. This is a challenge for researchers the world over.

The challenge is compounded for researchers in countries in the Global South. For instance, the pressure to publish in high-impact journals is a steeper barrier because those journals are predominantly in English and biased towards publishing data from the United States and Western Europe (Amano et al. 2016). With the exception of an emerging body of Chinese journals, local-language publications are broadly deemed lower tier – even those published in European-origin languages such as Spanish, Portuguese or French.

The metrics problem is further amplified for researchers who work on local challenges. Climate adaptation research is a case in point. Countries in the Global South are on the front lines of global warming, where context-appropriate adaptation strategies are crucial. These depend on highly localised data on complex factors such as weather patterns, biodiversity, community perspectives and political appetite. These data can be collected, curated, analysed and published by local researchers. In some cases, it is crucial that the work is done by them. They speak the necessary languages, understand customs and culture, are respected and trusted in communities and can thus access the traditional knowledge required to interpret historical change. This work helps to craft adaptations that make a real difference to people's lives. But it is also fundamental to high-level meta-research and analysis that is conducted later, far from the affected areas (Amano and Sutherland 2013).

Does the current evaluation approach scrutinise and give equal recognition to the local researcher who focuses on specifics and the researcher who generalises from afar? Does the current approach acknowledge that incentives are different for local and foreign researchers, and that those incentives affect research decisions? Are we adequately measuring and rewarding research that is locally grounded and globally relevant? In our view, the answer to all of these questions is no.

From no to yes

With the support and leadership of partners across the Global South, the IDRC decided to try something different. The result is a practical tool that we call Research Quality Plus (RQ+) (Ofir et al. 2016).

The tool recognises that scientific merit is necessary, but not sufficient. It acknowledges the crucial role of stakeholders and users in determining whether research is salient and legitimate. It focuses attention on how well scientists position their research for use, given the mounting understanding that uptake and influence begins during the research process, not only afterwards.

We think that the approach has merit beyond the development context. We hope that it can be tailored, tested and improved in a variety of disciplines and contexts, to suit the needs of other evaluators – funders such as ourselves, but also governments, think tanks, journals and universities, among others.

RQ+ has three tenets:

1. *Identify contextual factors.* There is much to learn from the environment in which research occurs. Instead of aiming to isolate research from how, where and why it was done, and by whom, evaluators should examine these contexts to reach a claim about quality. For the IDRC, this included five issues: political, data, research environments, the maturity of the scientific field and the degree to which a project includes a focus on capacity strengthening. For another funder, journal or think tank, these might – or should – be different.
2. *Articulate dimensions of quality.* The underlying values and objectives of the research effort need to be made explicit. Evaluators weigh these dimensions of quality using a formula that fits the context and goals of the research. The dimensions that matter to the IDRC are: scientific integrity (a measure of methodological rigour), legitimacy (a measure of the fidelity of the research to context and objectives), importance (a measure of relevance and originality) and positioning for use (the extent to which research is timely, actionable and well communicated).
3. *Use rubrics and evidence.* Assessments must be systematic, comparable and based on qualitative and quantitative empirical evidence, not just on the opinion of the evaluator – no matter how

expert they are. For the IDRC, this meant evaluators speaking to intended users, to others working in similar areas and to non-scientific beneficiary communities, as well as assessing research outputs and associated metrics.

Road test

The IDRC first used RQ+ in 2015. Independent specialists assessed 170 studies from seven areas of research the centre had funded in the previous five years. For each area, three specialists rated projects using the three tenets described, looking at empirical data for each study: bibliometrics, interviews with stakeholders and IDRC reports on the work. The reviewers decided independently what data to collect and compare for each project, and held panel discussions to reach a consensus on the final ratings for each project. More details are available in Ofir et al. (2016) and McLean and Sen (2019).

The RQ+ framework that embodied the three tenets for the IDRC (see Figure 1) encouraged a grounded, critical reflection on each project. And it helped systematic judgement to be applied across diverse contexts, disciplines and approaches to research. In exit interviews and follow-up discussions, the independent reviewers described the assessments as unlike any others they had done. They felt confident that the evaluation had been systematic, comprehensive and fair.

We learnt a lot from this process about the projects that the IDRC supports and how we could do better. For instance, we found that we need to prioritise gender across everything we fund, from climate modelling to the accessibility of justice, and not just in research projects that are aimed specifically at women and girls. As enshrined in one of the United Nations Sustainable Development Goals (SDG5), gender equality is key for unlocking development potential, so it was a dimension examined by the reviewers.

They found, for example, that a programme using national data sets to examine the implications of taxation and food labelling should have disaggregated the data by gender to achieve more with the same

Figure 1: The RQ+ Framework as used at IDRC

Framework components

The RQ+ Assessment Framework consists of three main components:

1. Contextual factors

Constraining and enabling contextual influences - within or external to the research effort - most likely to affect research performance are identified.

The categorisation of contextual factors using a rubric and a three point scale (e.g. low, medium, high) establishes a risk profile that is used to inform the quality assessment.

The contextual factors can be
1) constraining (negative) or
2) facilitating/enabling (positive)

Examples from IDRC experience:
1) Maturity of the research field
2) Research capacity strengthening
3) Risk in the data environment
4) Risk in the research environment
5) Risk in the political environment

2. Dimension and subdimensions

The four dimensions and their subdimensions encapsulate the quality assessment criteria.

Tailored for IDRC:
1. **Research integrity**
2. **Research legitimacy**
   2.1 Adressing potentially negative consequences
   2.2 Gender responsiveness
   2.3 Inclusiveness
   2.4 Engagement with local knowledge
3. **Research importance**
   3.1 Originality
   3.2 Relevance
4. **Positioning for use**
   4.1 Knowledge accesibility & sharing
   4.2 Timeliness and actionability

3. Evaluative rubrics

Performance is evaluated using customisable reseach quality rubrics.

Characterisation of each key influence, dimension and subdimension is done using tailored rubrics that combine quantitative and qualitative measures.

Ratings on an 8-point scale show four levels of performance (or progress). This is an example. Scales should be crated to fit a purpose or intention.

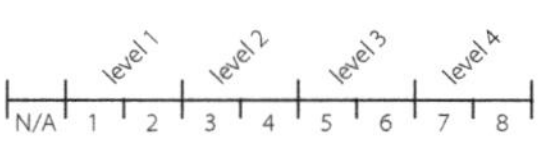

investment. Reviewers also highlighted exemplars, such as the African Doctoral Dissertation Research Fellowship programme, which helps PhD students to complete these at their home institutions, enabling greater uptake by female applicants who shoulder more family duties. The programme considers gender balance when selecting applicants, and in reviewing proposed research.

As a result, the IDRC has rolled out, among other things, a new data system to mine gender data and workshops for staff to share and see good work.

In our experience, conventional evaluations were never this challenging, but neither were they so motivating and useful.

Three myths busted

To draw more general lessons, the IDRC worked with an independent specialist to conduct a statistical meta-analysis using blinded data (see Gurevitch et al. 2018 for a description of the meta-analysis technique). We aggregated results from our seven independent evaluations of 170 components from 130 discretely funded research projects in natural and social science, undertaken in Africa, Asia, Latin America, the Caribbean and the Middle East (McLean and Sen 2018). This revealed three things.

Southern-only research is high quality. Research housed wholly in the Global South proved scientifically robust, legitimate, important and well positioned for use. Researchers in the region scored well across each of these criteria (higher, on average, than the Northern and North–South-partnered research in our sample). In other words, those most closely linked to a particular problem seem to be well placed to develop a solution. (See McLean and Sen 2019 for full results.)

This finding challenges assumptions that researchers in the north automatically strengthen the capacity of partners in the South (Bradley 2017). There are many positive reasons to support North–South research partnerships, but the data suggest that we must be strategic to optimise their impact.

Capacity strengthening and excellence go hand in hand. Too many funders assume that research efforts in which teams receive training and skills development inevitably produce poor-quality research. The meta-analysis found no such trade-off. In fact, we found a significant positive correlation between scientific rigour and capacity strengthening.

This suggests that research requiring a focus on capacity strengthening need not be avoided out of a desire for excellence. Indeed, it implies that the two can go hand in hand.

Research can be both rigorous and useful. In the fast-paced world of policy and practice, findings need to get to the right people at the right time, and in ways that they can use (see below 'Co-Producing climate adaptations in Peru'). We often hear of tension between sample saturation or trial recruitment and the decision-making cycle of policy-makers or industry implementers. Happily, the meta-analysis found a strong positive correlation between how rigorous research is and how well it is positioned for use.

This finding builds the case for investing in scientific integrity, in even the most applied and translational programmes.

Four concerns

We have four main concerns about RQ+ and how it can be refined and adapted for broader application.

First, bias is baked into our study. We used our own tool to examine research we had already supported. RQ+ focused our post-hoc evaluations on the values that matter to our organisation. The method examines our objectives and priorities, as we define them. Some would counter that it reifies them.

Second, this tool, much like all others, could have a distorting effect. For instance, by asking reviewers to examine integrity and legitimacy – issues that we identify as fundamental to our success – we turned their attention away from other factors, such as productivity (volume of publications and outputs) and cost efficiency.

Third, there is the risk that RQ+ results become isolated if they are not comparable with the prevailing measures of research quality used by the global research enterprise. Is RQ+ just another demanding hurdle for researchers in the Global South? That's a question we are still working to answer.

Fourth, RQ+ costs more and takes longer than asking two or three peers to offer their opinions. Our hunch is that it takes almost twice as much time and money, largely because it requires empirical data collection by the evaluators. For us, that is time and money well spent: the results help us to hone our approach to funding and engagement.

These concerns will guide our efforts to improve RQ+, as will input from our peers and partners.

CASE STUDY: Co-producing climate adaptations in Peru

More than 500 000 people live in the Mantaro Valley in central Peru, where agriculture is the main source of income. The valley's small-scale farmers provide most of the vegetables and grains consumed in the capital, Lima, but are struggling to respond to the increasing frequency and intensity of extreme droughts, heavy rainfalls and frosts.

Using new and creative combinations of physical measurements and participatory engagement methods such as community mapping, the Geophysical Institute of Peru in Lima is providing a clearer picture of how the climate has changed in the region. This research is informing local policy and guiding adaptation actions. The project mapped hotspots across the region that were susceptible to climate change, and convened discussions with farmers and fishers about how they could adapt schedules and techniques to minimise its impact.

The team did not rush to publish the research in top-tier Western journals, partly because of the English-language barrier but largely because of the urgency of the problem. The research outputs needed to be immediately understandable and usable, so the team rapidly published its findings in working papers and reports (many of which were collected in a Spanish-language book (IGP [vols 1 and 2] 2012). These were immediately accessible to those in local government who needed the evidence to steer the response. As such, predominant metrics do not capture the value of this work.

The RQ+ review shone a different light on this project and its achievements. It scored highly for integrity (including innovative blending of techniques for knowing the climate), for being legitimately grounded in local needs and knowledge, for addressing an urgent problem, and for focusing on uptake and action.

More like this

What next? If the trillions of dollars being invested in research globally each year (R&D Magazine 2017) are to make a difference, we must do better than crude quantification of citations, as the *Leiden Manifesto* (Hicks et al. 2015) and the *San Francisco Declaration on Research Assessment* (ASCB 2012) have made clear.

We believe RQ+ presents a practical solution. The approach and findings of our meta-analysis now need replication in other contexts. At IDRC, we are planning another retrospective assessment in 2020. We are excited by what progress and shifts it might uncover. We are already looking at ways we can use RQ+ for grant selection, monitoring the progress of individual projects and communicating our organisational objectives to funding partners and applicants.

Similarly, we encourage other funders and institutions to improve their evaluations in three ways: consider research in context; accept a multidimensional view of quality; and be systematic and empirical about evidence collection and appraisal. It's time science turned its greatest strengths on itself – experiment, appraise, debate and then improve.

Notes

1. This chapter was originally published in *Nature* as, 'A better measure of research from the Global South', Lebel and McLean (2018).

References

Amano T and Sutherland WJ (2013) Four barriers to the global understanding of biodiversity conservation: Wealth, language, geographical location and security. *Proc. R. Soc. B* 280: 20122649

Amano T, González-Varo JP and Sutherland WJ (2016) Languages are still a major barrier to global science. *PLoS Biol.* 14: e2000933

ASCB (2012) Annual Meeting of the American Society for Cell Biology. *San Francisco Declaration on Research Assessment*. ASCB

Bradley M (2017) Practical action. In: LJA Mougeot (ed.) *Putting Knowledge to Work*. Ottawa, Canada: IDRC

CIHR (Canadian Institutes of Health Research) (2013) *Evaluation of CIHR's Knowledge Translation Funding Program*. http://www.cihr-irsc.gc.ca/e/47332.html

Gurevitch J, Koricheva J, Nakagawa S and Stewart G (2018) Meta-analysis and the science of research synthesis. *Nature* 555: 175–182

Hicks D, Wouters P, Waltman L, de Rijcke S and Rafols I (2015) Bibliometrics: The Leiden Manifesto for research metrics. *Nature* 520: 429–431

Holmes B (2016, 27 April) The rise of the impact agenda. Paper presented at Fuse International Conference on Knowledge Exchange in Public Health, Newcastle, UK. https://drive.google.com/file/d/1rjfOo7hF9or1Rdd0Tl0lp4-98ZWNuTKP/view

IGP (Instituto Geofisico del Peru) (2012) *Eventos meteorológicos extremos (sequías, heladas y lluvias intensas) en el valle del Mantaro*. Vol. 1

IGP (Instituto Geofisico del Peru) (2012) *Manejo de riesgos de desastres ante eventos meteorológicos extremos en el valle del Mantaro*. Vol. 2

Lebel J and McLean R (2018) A better measure of research from the Global South. *Nature* 559: 23–26

McLean RKD and Sen K (2018) *Making a Difference in the Real World? A Meta-Analysis of Research for Development*. Ottawa, Canada: IDRC

McLean RKD and Sen K (2019) Making a difference in the real world? A meta-analysis of the quality of use-oriented research using the Research Quality Plus approach. *Research Evaluation* 28(2): 123–135

Ofir Z, Schwandt T, Duggan C and McLean R (2016) *Research Quality Plus (RQ+): A Holistic Approach to Evaluating Research*. Ottawa, Canada: IDRC

R&D Magazine (2017) Global R&D funding forecast. Winter: 3–6

Wilsdon J, Allen L, Belfiore, E, Campbell P, Curry S, Hill S et al. (2015) *The Metric Tide: Report of the Independent Review of the Role of Metrics in Research Assessment and Management*. Report. Higher Education Funding Council of England

Call to action: Transforming 'excellence' for the Global South and beyond

Erika Kraemer-Mbula, Robert Tijssen, Matthew L. Wallace, Robert McLean, Liz Allen, Rodolfo Barrere, Joanna Chataway, Diego Chavarro, Chux Daniels, Jean Lebel, Elizabeth Marincola, Enrique Mendizabal, Cameron Neylon, Annette Ouattara, Falak Raza, Yaya Sangaré, Suneeta Singh, Fajri Siregar, Vincent A. Ssembatya and Judith Sutz

This book highlights gaps and shortcomings in how the notion of 'excellence' is currently applied across research ecosystems. It argues that we must do better if scientific research is to fulfil its promise – as a productive force in creating a healthier, happier, more prosperous society, in particular in the Global South, where the hazards of striving for 'excellence' can lead to troubling effects. It is time for change, and this book highlights ideas for how we can achieve this.

From a range of theoretical and practical perspectives, we dug deep to understand the current scope of problems associated with the general notion of 'research excellence', especially in the context of performance assessment systems applied by funders. We have identified deficiencies in current systems of research evaluation that have the potential to further exacerbate the gaps between North and South. We have proposed new ideas, informed by knowledge and experiences working across the Global South, that offer alternatives to the status quo.

But the discussion contained in this book is not just about research in low- and middle-income countries (LMICs). Positive change comes from fresh thinking, and the Southern research community has no shortage of this resource. Indeed, with more than half of the world's population growth between now and 2050 to come from LMICs, the entire globe will depend on this burgeoning talent pool for the knowledge and innovation a prosperous global future will need to tap into. Accordingly, this book has laid the groundwork for a new vision of what is important for high-quality and high-impact research to emerge. There is a clear appetite among researchers, funders and administrators to have research excellence better reflect context and the ultimate objectives of science policies and research initiatives, reconciling the needs of scientists and society at large.

The misuse of the term 'research excellence' has led to debates around the globe. There is an opportunity for new ideas from the Global South to lead to positive change not only in their respective research ecosystems, but also around the globe. At the heart of the issue is the need for a pluralistic view of what quality means and a better understanding of what it means to recognise the 'best' researchers, as well as a drive to operationalise and systematise our knowledge on the issue. Simplistic views of 'excellence' in scientific output are unhelpful in a world that enables research to be shared in increasingly more open, accessible and usable forms. While avoiding its most egregious misuses, we can also reclaim the term 'research excellence' by building improved or radically new assessment tools and science policies, with more appropriate stakeholder expectations based on norms and values that align with research practices and goals of the Global South.

'Beyond buzzwords': Research excellence should not be taken for granted, but made transparent, precise and tailored to context, or avoided altogether

Research *excellence*, while incorporating the ideal standards (whatever they may be) of 'high-quality' science, is fundamentally different from research *quality* in that it implies superiority and a scientific 'elite'. In addition, as many of the contributions to this collection have highlighted, it has become a powerful rhetorical technique, particularly among funders and institutions. While references to 'top' researchers, institutions, papers, etc., may not be problematic *per se*, excellence as a buzzword or public relations tool has become disproportionately dominant. At best, the term provides very little information on the science – or scientists – that it qualifies, nor does it say much about the potential use, re-use or practical application of research. At worst, it can lead to perverse incentives and introduce significant biases in how research from the Global South is judged. There are advantages and drawbacks to concentrating rewards and resources among a small group of extraordinary 'excellent' researchers, particularly where resources are scarce. Policies and institutional strategies should be able to choose to eschew the term 'excellence' not to decrease the quality of the research performed, but to focus efforts on strengthening research ecosystems or focusing on specific societal challenges, for instance.

Transparency means being open and systematic about how we approach the definition and measurement of research quality or excellence. Advancing the quality of research will require quantitative and qualitative approaches that are tied overtly to the underlying objectives of the work. But to what degree, and how, is excellence measurable? Meaningful research evaluation must be purpose-built. It cannot simply be transposed or assumed from other tools, or from political discourses, and the effects of evaluation frameworks, particularly 'excellence'-driven, must be explicitly considered. Evaluators must reflect on the intentions, and potential unintended consequences, of their efforts.

Finally, more efforts should be devoted to measuring meaningful research impact, which is perhaps distinct from the notion of 'research excellence' that currently prevails, but should be an increasingly important approach to research evaluation. It is time for funders, universities, governments and others to innovate in tailoring the processes of reviewing research proposals, setting up incentive structures, and gauging the results of research projects. Addressing research funding and publication processes is especially critical for achieving this type of change. We need to recognise, describe and incentivise research that has value across a variety of local, national and global contexts; it needs to be done well, be valid, but need not be 'excellent' or 'superior'.

'Many voices': Excellence is pluralistic, and should be used to recognise diverse forms of scholarship

To reinforce this last point, research has value and importance in different contexts, places and in time. Where the term 'excellence' is used, it must be seen as fundamentally pluralistic. There is no globally accepted definition of excellent science, and evaluators should accept the opportunity for exploration and contextualisation this freedom provides. We need to move away from a homogenous view of both research quality and research excellence to allow for science to be measured against local priorities or the critical needs of national research ecosystems.

Second, a pluralistic view of research excellence is intertwined with the diversity in the knowledge that is produced through research. Scientific results are produced in specific settings, with specific values, objectives and institutions guiding the work. We must enable different forms of high-quality knowledge to be produced through different methodologies and in different languages and formats. This not only helps develop a multiplicity of better-tailored standards for assessing research in different contexts, but can also help research and researchers from the Global South be better recognised locally and globally, rather than being restricted to a narrow range of 'Northern' indicators and metrics.

Accepting pluralism also links to being purposeful and transparent about how terms such as 'research excellence' are used. Research evaluations can and should have different objectives. At times, evaluations should seek to reward the top performers; at other times, evaluations should aim to shed light on novel or breakthrough ideas; and sometimes, evaluation should be used to prioritise research that addresses pressing societal or environmental challenges.

'Towards operationalisation': Actors and platforms that can change how science is done

Meaningful change will require a large-scale systematic effort. Structural change is needed. Many contributing actors – such as researchers, funders, universities and journals, to name but a few – play particular roles in valuing and assessing research. With an all-of-system understanding of the issue, different actors should consider how their efforts can make a difference for their own community, and how change may contribute to wider systems' transformation. Here there is a significant opportunity. New partnerships and platforms that rest on the collective action of multiple actors have the potential to stimulate change in deep and far-stretching ways. For example, new publishing platforms and evaluation schemes can help value locally relevant knowledge and move beyond a 'catch-up' mentality – this is seen through 'open science' leadership in Latin America, for instance. Another example is the African Academy of Science which, in collaboration with key international donors such as the Wellcome Trust, has developed tools and programmes to 'shift the centre of gravity' of global research. And national granting councils are increasingly at the forefront of these efforts. The Science Granting Councils Initiative (SGCI) in sub-Saharan Africa contributes to empowering national research agencies through dialogues and capacity building to focus precisely on collaboratively operationalising new ideas and improving the effectiveness of grant-making in contexts where funds and other resources are scarce.

Many theoretical underpinnings, methodologies and performance indicators are there, as evidenced by the contributions in this book. Now is the time for dedicated leadership in operationalising, adapting and continuously improving on them, with a view to either moving beyond or to reclaiming 'research excellence' in the Global South and globally. We need compelling, effective, affordable, scalable and sustainable solutions. This will have important 'bottom-up' and 'top-down' implications on how different modalities of knowledge are perceived and produced, shared and used, and on researchers' careers, as high-quality science and scientists are increasingly called upon to tackle the most pressing socio-economic and environmental problems at national, regional and global scales.

About the authors

Liz Allen is director of Strategic Initiatives at F1000 and is involved in shaping new initiatives and partnerships to promote and foster open and sustainable models of scholarly publishing. Liz has a background in research assessment and evaluation. Prior to joining F1000 in 2015, she spent over a decade leading the Evaluation Team at the Wellcome Trust. Liz is currently a visiting senior research fellow in the Policy Institute at King's College London, with a particular interest in science policy research and research-related indicators, is a board director of Crossref, and serves on the Advisory Board for the UK Software Sustainability Institute.

Rodolfo Barrere acts as Ibero-American Network on Science and Technology Indicators (RICYT) coordinator. He holds a PhD in social sciences, with his thesis on the dynamics and evolution of the production of scientific and technological information. Throughout his career he has dedicated his work to production, management and analysis of scientific, technological and innovation information. Within RICYT, he has focused on the characteristics of indicators production in Ibero-American countries. After several years of work at CAICYT, CONICET's documentation institute, he developed broad experience in the production of bibliometric indicators. He has taken part in several research projects funded by OECD, UNESCO, IADB, World Bank and the European Union.

Joanna Chataway is professor of science and technology policy and head of department of Science, Technology, Engineering and Public Policy (STEaPP) at University College London (UCL). STEaPP is in the Faculty of Engineering and reflects Joanna's commitment to interdisciplinary and transdisciplinary research and academic work. She has worked for many years in science and innovation policy and at the intersect of academic and policy research. She has held senior positions at the Science Policy Research Unit (SPRU), University of Sussex, RAND Europe and The Open

University. Her research spans high-, middle- and low-income country contexts and includes work on health research and innovation policy.

Diego Chavarro holds a doctorate degree in science, technology and innovation policy from the Science and Policy Research Unit, University of Sussex. Diego is a researcher with an interest in science, technology, and innovation policy, with a particular interest in how scientific knowledge is valued in society. Specifically, he studies the evaluation of research and researchers in non-dominant contexts: communities who use languages other than English for scientific publishing, geographies not considered economic powers, disciplines that have a lower status than the natural sciences, among others. Diego has worked for a range of organisations in the academic, public and civil society domains. This has allowed him to learn their different perspectives on research evaluation. In his work, he puts these perspectives into dialogue to gain a more comprehensive understanding of evaluation practices and suggest how research policies can be improved, using evaluation to develop research capacity.

Chux Daniels is a research fellow in science, technology and innovation (ST&I) policy at the Science Policy Research Unit (SPRU), University of Sussex. He holds a doctorate in science and technology policy studies from SPRU, University of Sussex. His research connects ST&I and public policies in ways that contribute to addressing development challenges and fostering transformative change across sectors, systems and societies. His areas of research interest include ST&I, public policies and policy processes, capabilities, policy-making (formulation, implementation, evaluation and governance), research excellence, inclusion in ST&I, entrepreneurship, ST&I indicators and metrics, and technology management. He leads the Transformative Innovation Policy (TIP) Africa hub research project, which involves Ghana, Kenya, Senegal and South Africa.

Erika Kraemer-Mbula holds the DST/NRF/Newton Fund Trilateral Research Chair in Transformative Innovation, the 4th Industrial Revolution and Sustainable Development. She specialises in science, technology and innovation policy analysis and innovation systems in connection with equitable and inclusive development. Initially trained as an economist, she holds a master's degree in science and technology policy from the Science and Policy Research Unit at the University of Sussex, and a doctorate in development studies from the University of Oxford. She is currently a research associate at the Centre for Law, Technology and Society, University of Ottawa (Canada), the Institute for Economic Research on Innovation (IERI) at Tshwane University of Technology (South Africa) and a researcher at the DST-NRF Centre of Excellence in Scientometrics and Science, Technology and Innovation Policy (SciSTIP) (South Africa).

Jean Lebel was appointed president of Canada's International Development Research Centre (IDRC) in 2013. As president, Jean leads the centre's contributions to Canada's development, foreign policy and global innovation efforts. He is responsible for significant funding partnerships with Canadian and foreign governments, philanthropic organisations and the private sector. Jean previously served as vice-president, Program and Partnership Branch, overseeing all IDRC programming, as well as director, Agriculture and Environment. He holds a doctorate in environmental sciences from Université du Québec à Montréal and an MScA in occupational health sciences from McGill University.

Elizabeth Marincola is senior advisor for science communications and advocacy at the African Academy of Sciences, which drives scientific research across Africa. Elizabeth is an international leader in open access publishing, science advocacy, communications, education and public engagement. She was CEO of the open access publisher PLOS, after serving as a long-time PLOS board member. She launched *AAS Open Research*, an innovative scholarly publication. Elizabeth was president of the Society for Science & the Public, publisher of *Science News* magazine, and executive director of the American Society for Cell Biology and the Coalition for Life Sciences. She served on the founding boards of PubMed Central and eLife, and on numerous US and EU advisory commissions on open science. She is currently on the board of the Johns Hopkins Bloomberg School of Public Health Center for Humanitarian Health. She received her bachelors and MBA degrees from Stanford.

Robert McLean is senior programme specialist in policy and evaluation at Canada's International Development Research Centre (IDRC). He is concurrently a research fellow in the Integrated Knowledge Translation Research Network (IKTRN) based at the Ottawa Hospital Research Institute/University of Ottawa, where he leads research looking at the role funders/donors play in turning innovation into action. Rob's broad interests lie in understanding how human creativity might help to create a better world. He has worked across the academic, government, private and NGO sectors. He has published these experiences and his work in venues ranging from *Nature* to the Stanford Social Innovation Review. Rob is a PhD candidate in the Department of Medicine and Health Sciences at Stellenbosch University, South Africa. He holds an MSc from the University of Manchester, England, and two undergraduate degrees, following studies at Carleton University, Canada and the University of KwaZulu-Natal, South Africa.

Enrique Mendizabal is the founder and director of On Think Tanks (OTT). He is a research affiliate at Universidad del Pacífico, as well as fellow and international member of the fellowship council of the Royal Society of Arts. Before founding OTT, he worked for the Overseas Development Institute (ODI) where he headed the organisation's research into how, and why, research informs development policy. At ODI he co-founded the Outcome Mapping Learning Community and the Evidence-based Policy in Development Network. Enrique is the co-founder of Politics & Ideas, the Peruvian

Alliance for the Use of Evidence, the PODER Award to the Think Tank of the Year in Peru, and Evidence Week in Latin America. Enrique's work, across these efforts, focuses on promoting the generation, communication and use of evidence in public policy.

Cameron Neylon is professor of research communication at the Centre for Culture and Technology at Curtin University, where he is project lead on the Curtin Open Knowledge Initiative. He is also director of KU Research, and an advocate of open research practice, who has worked in research and support areas including chemistry, advocacy, policy, technology, publishing, political economy and cultural studies. He was a contributor to the Panton Principles for Open Data, the Principles for Open Scholarly Infrastructure, the altmetrics manifesto, a founding board member and past president of FORCE11 and served on the boards and advisory boards of organisations including Impact Story, Crossref, altmetric.com, OpenAIRE, the LSE Impact Blog and various editorial boards. His previous positions include advocacy director at PLOS, senior scientist (Biological Sciences) at the STFC and tenured faculty at the University of Southampton. Alongside his earlier work in structural biology and biophysics, his research and writing focuses on the culture of researchers, the political economy of research institutions and how these interact, and collide with, the changing technology environment.

Annette Lhaur-Yaigaiba Ouattara holds a doctorate in sociology from Université Félix Houphouet-Boigny. She is a senior lecturer at the Université Nangui Abrogoua (Abidjan, Côte d'Ivoire) and an associate researcher at the Centre Suisse de Recherches Scientifiques in Côte d'Ivoire. Since 2008, she has been working at the Programme d'Appui Stratégique à la Recherche Scientifique (PASRES) in Côte d'Ivoire, where she was first the assistant to the executive secretary, then in charge of capacity building and partnerships. PASRES is a collaborative Swiss-Ivorian programme that has become one of the main structures for supporting research and innovation in Côte d'Ivoire, since its inception in 2007.

Falak Raza holds a postgraduate degree in development studies. She has worked with think tanks and development consultancies, and contributed to several research studies in South Asia, focused on accessing elementary education and school choice, maternal and child health and nutrition, marginalisation and social exclusion vis-à-vis rights and entitlements, and women's safety and security in rural public spaces. A qualitative researcher by choice, she has experience in implementing qualitative methodologies and developing participatory tools, and has published independent articles and opinion pieces on some key issues grappling the Indian subcontinent. Presently, Falak is engaged with the Research Compliance department at the International Center for Research on Women, a global non-profit organisation, where her work has helped set ethical standards for conducting human subject research, keeping in mind the contextualities of the Global South, and with particular attention to research with disadvantaged and vulnerable groups.

Yaya Sangaré has expertise in two fields: materials science and organisational management. Following the completion of a doctorate in materials physics from Université de Montpellier (France) and graduate degrees from the Institut d'Administration des Entreprises de Poitiers (France), he taught at the Institut National Polytechnique in Yamoussoukro (Côte d'Ivoire) and later joined the Chamber of Commerce and Industry as the director of training and development for executives. He then held various positions in the private sector in Côte d'Ivoire and Burkina Faso. This career path provided him with critical insight into the expectations of industry and allowed him to pursue efforts to build linkages with the research and innovation sectors. Since 2007, Yaya has been the executive secretary of the Programme d'Appui Stratégique à la Recherche Scientifique (PASRES), one of the main structures for supporting research and innovation in Côte d'Ivoire.

Suneeta Singh (MD, DCH) is a medical doctor with postgraduate qualifications in paediatrics and public health from the Lady Hardinge Medical College, India. During more than 30 years in the development sector, she has worked in academics, in bilateral and multilateral funding organisations, and established a research and consulting firm, Amaltas. The Delhi-based Amaltas Consulting Pvt. Ltd. is devoted to developing intellectual capital to accelerate improvements in the lives of people. The organisation's work on more than almost 70 projects in the past decade has helped bring about programmatic and policy changes in developing countries. A key area of concentration has been research quality and research systems. Suneeta has worked with research funders to study the quality of their projects and portfolios. Her advice has been sought on improvements that could be made in design and implementation, as well as to translate research findings into policy and programmes.

Fajri Siregar is a lecturer at the University of Indonesia and a former director of the Centre for Innovation Policy and Governance (CIPG, Jakarta, Indonesia). Fajri has acted as a consultant for Indonesia's Ministry of Research and Higher Education and the Knowledge Sector Initiative (KSI) on various occasions. He also co-authored the report *Reforming Research in Indonesia: Policies and Practices* in 2016. Fajri has a keen interest in the production of knowledge, particularly social science. He has a wide experience of empirical research, including research on media policy, the use of ICT for good governance, open government and the growth of creative economy. He is currently undertaking a PhD in anthropology at the University of Amsterdam, with a dissertation on knowledge production by NGOs in post-Suharto Indonesia.

Vincent A. Ssembatya is currently the director of quality assurance at Makerere University in Uganda. He was dean for the Faculty of Science at Makerere University for the period 2005 to 2009. Vincent is a member of the Governing Council of the African Network for Scientific and Technological Institutions (ANSTI) representing the East African Region. He is also an associate of the Centre for Higher Education Transformation (CHET) based in Cape Town, South Africa. He has co-authored scholarly works on higher education, as well as pure mathematics. He holds a PhD in mathematics from the University of Florida.

Judith Sutz is a professor at the University of the Republic, Uruguay, where she teaches science, technology, innovation and development. She is the academic coordinator of the University Research Council, leading a research group that deals with the design and implementation of a set of competitive funds schemes aimed at fostering university research. Research evaluation is a core issue for this group, both from an ideological and theoretical perspective and as an empirical field of practice and analysis. Two recent papers on the matter are Bianco M, Gras N and Sutz J (2016) Academic evaluation: Universal instrument? Tool for development? *Minerva* 54(4): 399–421 and Arocena R, Goransson B and Sutz J (2019) Towards making research evaluation more compatible with developmental goals. *Science and Public Policy* 46(2): 210–218.

Robert Tijssen straddles the Global South and Global North. He holds the Chair of Science and Innovation Studies at Leiden University (Netherlands), but is also part-time full professor – since 2015 – at the Centre for Research on Evaluation, Science and Technology (Stellenbosch University, South Africa) and affiliated to South Africa's DST-NRF Centre of Excellence in Scientometrics and Science, Technology and Innovation Policy. Robert's active involvement in 'research excellence' dates back more than 15 years, when he co-developed the 'top 10% most highly cited' indicator of 'international scientific excellence' (2002), as well as introducing the idea of 'scoreboards of research excellence' (2003), both specifically designed for Global North applications. Being active in South Africa since 2005, his interests and academic work on 'excellence' have gradually broadened and shifted towards addressing Global South issues and problems – not only with regard to its conceptualisation, but also its application in contextualised evidence-based evaluations of research performance.

Matthew L. Wallace has been working in the science policy field for over a decade. Currently, as a senior programme specialist at the International Development Research Centre (IDRC) in Ottawa, Canada, his focus is on issues broadly related to science systems, including engineering education, science advice to governments, granting agencies, industrial research collaboration and careers of scientists. His previous research in the field has focused on indicators for science, research portfolio management, the history of scientific disciplines and priority setting in research. Matthew has masters degrees in physics (Ottawa) and in science and technology studies (Strasbourg), as well as a PhD (UQAM) in the history of science. He previously worked as a science policy advisor and a senior evaluator at a Canadian federal ministry and granting agency. Developing, applying and questioning definitions of research quality and excellence have been common threads during his career as a scholar, evaluator and policy-maker.

Index

Page numbers in *italics* indicate figures and tables.
The letter 'n' after a page number indicates a note.

accountability v, vi, 2, 63, *244*
African Academy of Sciences Open Research 41, 241–243, 263
African Doctoral Dissertation Research Fellowship programme 253
African scientists 84–85, 241
African Union (AU) 48, 87
African universities 83–84, 161, 162
  see also Makerere University (Uganda); universities
agendas *see* research agendas
Altbach, P. 21
Amaltas Consulting 165, 166, 173, 176n2
applied science/research
  versus basic science/research 50
  and measurement of RE 71
  in Stokes Typology of Research *166*
  see also entries under science
Argentina 22, 26, 30, 195, 221, 227
Arocena, R. 25, 83
arXiv 239, *240*
ASEAN countries 124, *125*
assessment/measurement of RE
  broader perspectives on 73–76, 260
  and debates around excellence 4–10
  and gender disparities in performance 80–82
  in Indonesia 123–126, 127–132, *133*, 134–135
  need for transparency 261
  in neo-colonial context 97, 98
  in open access publishing 240, 244–245
  related to values and motivations 12–13
  and research funding 27–28
  responsible assessment of 60–62, 63, 71–73
  and Southern views on RE 167–171, 175–176
  and unit of assessment 65, 72
  see also bibliometric measures; citation counts; evaluation of

research; outputs of research; peer-review processes; publication counts/output; quantitative indicators
Australia 94, 95, 113n8, 122, 175, 204
autonomy of research
critical in developing/transitional countries 97
deployment of RE in defence of 95–96
and 'impact agenda' 95
linked to peer-review processes 41, 43, 95
Polanyi's views on 41
power structures related to 44
and relevant/embedded research 41, 42, 46, 47
in ST&I policy 206
and tensions within SGCs 41, 42–48
in think tanks *187*
awards and prizes 7, 59, 67, 145, 237, 239

Baertl, Andrea 186, *187*
Balkan countries 182, 192
Barkhuizen, N. 84
basic science/research
vs applied science/research 50
in Stokes Typology of Research *166*
see also entries under science
Bezuidenhout, A. 84
bias in research
and gender disparities 89
in peer-review processes 3, 234
and publication of research *244*, 250
in RQ+ model 255
towards researchers 3, 173
bibliometric measures
overview of 222–226
and debates around excellence 6–8
as useful standard for knowledge production 220, 235
see also assessment/measurement of RE; citation counts; outputs of research; publication counts/output; quantitative indicators
bioRxiv 239, *240*
Botana, N. *see* Sabato-Botana Triangle
branding 191
Brazil
open access publishing 110–111, 111
R&D expenditure *125*, 221
research evaluation system 204, 210–211
support for scientific journals 225
universities' share in SCOPUS-indexed articles 226–227
BRICS countries 124, *125*
Burnett, Keith 194

Calvert, J. 50
Cameroon 85
capacity building
challenges in developing countries 104
correlation with RE/rigour 254
in PASRES programme 144–145
in SGCs 50–51
and Southern views on RE 176n4
in Uganda 159
via partnerships/collaborations 53
capital, forms of
and bias towards established researchers 173
and gender disparities in research performance 82
required by SGCs 41, 45, 47, 48, 51
Carden, Fred 185
Carli, G. 68
Ceci, S.J. 82
Center for the Implementation of Public Policies for Equity and Growth 195
centres of excellence 59, 66, 127–128
Chatham House 194

Chile
research evaluation system 210–211
ST&I indicators 28
support for scientific journals 225
think tanks 180, 195–196
universities' share in SCOPUS-indexed articles 226–227
China *125*, 180
Cilliers, F.V. 84
citation counts
and bias towards established researchers 173
as criterion for evaluation 40, 61, 66
as crude quantification 256
and debates around excellence 6, 7, 96
fears of misuse of 71, 72
and gender disparities 81, 84
and language of researchers 109, 172–173
limitations of dominant evaluation approaches 249–250
and open access publishing 244
and Southern views on RE 167, *168*, 170
as Uruguayan NSR criterion 35
use in Mexico 19
used as indicator in RES 204, 206, 212, 215
see also assessment/measurement of RE; bibliometric measures; outputs of research; publication counts/output; quantitative indicators
climate change 250, 256
cOAlition S 236, 241
collaboration
as evaluation principle in sustainable development 209, 212, 213, 214–215
and raising awareness of research quality 135
as sub-theme of productivity 125
Colombia
research evaluation system 204, 210–214, 215
science-funding strategies 49, 53
universities' share in SCOPUS-indexed articles 226–227
colonialism 93, 94, 97, 98–99, 104
see also decolonisation; neo-colonial agenda of RE; post-colonial countries
communication strategies (think tanks) 190–197
community engagement 109, 188–189
see also engagement in research
conferences
in Indonesia *133*
organised by PASRES 145
confidence-building 96–97
Confraria, H. 67
Connell, R. 105
consultation processes 74
costs, research
of publishing 234–235
and RQ+ model 255
Côte d'Ivoire
and journal publications 143–144
and quality of research 139
see also PASRES programme (Côte d'Ivoire)
Court, Julius 185
credibility, and think tanks 185–190, 197–198

Dahler-Larsen, Peter 109
Darlington, Richard 193, *193*
data
analysis 169, *169*
collection 71, 169, *169*, 186, 255
management 111, 186
manipulation 8
sharing 62, 111

databases
in Indonesia 132
and language of researchers 3
SCOPUS 129, 159, *160*, 223, 224, *225*, 227, 239
and searches for articles/material 239
Web of Science 223, 224, 239
decolonisation 45, 93, 98, 106
see also colonialism; neo-colonial agenda of RE; post-colonial countries
Department for International Development (DFID) 43, 55n8, *158*
design, research 169, *169*, 176n3, 176n4
developing countries *see* low- and middle-income (developing) countries (LMICs)
development, and Sabato-Botana triangle 22–23, *24*, 99–103, 108
developmental universities 24–25, 83
directionality principle 209, 212, 213, 214–215
dissemination of research
and changing publishing landscape 238–239
frustration with legacy publishers 233–234
pre-internet manual procedures for 235
and Southern perceptions of funders' priorities 176n3
and Southern views on RE 167, *168*
diversity
and social knowledge creation 108
see also gender equality: gender diversity
Donovan, C. 170, 172–173

economic capital 45, 47, 48
see also capital, forms of
economic growth and development 44, 119–120, 206, 207, 221, 226
Egypt 248
embedded excellence
gains for researchers and funders 43–44
new approaches for 54–55
and relationship with autonomy 42, 46, 47
see also relevance of research; research excellence (RE)
engagement in research
and assessment of proposals 32–34
community engagement 109, 188–189, *189*
component of RQ+ model *253*
as 'damaging' to RE 92–93, 95
importance of variety of 'capitals' 45, 47, 51
and inclusion of 'non-scientists' 48–49
indicators for academic engagement 226–230
and issues around autonomy 44–45
as part of inter- and transdisciplinary exercises 52–53
and participation principle 209, 212, 213, 214–215
public engagement 191–194, 196–197
and relevance of research 51–52, 53
and Southern perceptions of funders' priorities 176n3
and Southern views on RE 167, *168*
engineers (Indonesia) 124–125, *126*
English language, reporting/publishing in 3, 21, 72, 103, 109, 170, 171–173, 175, 250
Enhancement of the Quality of Research programme (Uruguay) 32
enrolments (Uganda) 147–148, 150, 152–153
see also Makerere University (Uganda)

environment, research 68, 251, *253*
ethics in research 61, 62, 169, *169*, 174, 176n3, 240, *244*
European Union (EU) 141, *158*, 226
evaluation of research
in colonial settings 94–95
and distinction between 'excellence' and 'impact' 92–93
locally based research 250
meta-analysis of research projects 254–255
by NSR (Uruguay) 34–38
and research outputs 61
and Southern views on RE 169–170, *169*
theories underpinning evaluations 203–204
value of pluralistic vision of RE 11
and zero-sum/threshold RE 69–70, 75
see also assessment/measurement of RE; peer-review processes; research evaluation systems (RES); Research Quality Plus (RQ+) model
excellence, definitions of 4, 59
see also embedded excellence; research excellence (RE)
Excellence Initiative (Germany) 20

F1000 platforms 41, 239, 240, *240*, 242, 243
FAIR data/software 243, *244*
Feller, I. 89
Ferretti, F. 5, 26, 68, 112n7
Flores, Walter 188–189
France
French language 171–172, *172*
universities 20
fraud 8, 93
Fudickar, R. 67

funding of research
and concept of excellence 261
and distinction between applied and basic science 50
in favour of vested interests 69
and gender issues 79, 84
in Global South/LMICs context 62–64, 157
in Indonesia 132
must enable embeddedness 44
need for clear assessment criteria 27–28
and open access publishing 236, 240
in PASRES programme 140–141, 143
post-1945 expansion of 94
for R&D 123, *125*, 221, 224, *225*
and SGCs 44–45, 46, 48, 49, 50, 51–52, 52, 53
and Southern perceptions of funders' priorities 169, 176n3
in sustainable development context 209–210, 212, 213
for think tanks 195
in Uganda 152, 157–161, 162
and ultimate value of research 237

Garfield, Eugene 205
gender equality
and disparities in research performance 80–82
and gender diversity 79–80, 86–89, 105
and gender gap 79, 80
in Global South context 83–86, 87, 88
and RQ+ model 252–253, *253*
gender sensitisation 168–169
GERD (Gross Expenditure on Research and Development) 123, 124, *124*, *125*
German universities 20
Gläser, J. 66

Global Innovation Index 123, 123–124
Global North
development of bibliometric approaches 7
and dominance of standards 21, 26, 74, 98, 105–106, 165, 167
practices emulated by LMICs 62
and quality of research 214, 254
Global Research Council 88
Global South
barriers to publishing 233, 241, 250
bias toward researchers 3
clarification of the term 4
gender-related issues 83–86, 87, 88
and multi/inter/transdisciplinary work 171, 174
open access publishing 110–111, 241–243
and quality of research 214, 254
and RE, assessment of 74–76
and RE, funding issues 62–64
and RE, implications of embracing 10–13
and RE, Southern perspectives on 165–173, 173–176
taking a global lead in research 110–112
visibility of research(ers) 11, 72
see also locally based research; low- and middle-income (developing) countries (LMICs); names of countries; Sub-Saharan Africa (SSA)
global warming 250, 256
grants, research 67, 84, 139, 176, 239
granularity 72
Graversen, E. 66
grey literature 171
Gross Expenditure on Research and Development (GERD) 123, 124, *124*, *125*
Guatemala 188, *189*, 197
H-indices 74, 96, 130
Hazelkorn, E. 19–20
health and well-being 160, 166, 173, 237
Hess, D. 21–22
Hicks, D. 71
higher education sector
in Africa 161, 161–162
in Uganda 148, 148–149, 150, 151–152, 153–154, 161, 162
see also Makerere University (Uganda); universities; university rankings
Hottenrott, H. 67
Hunter, L.A. 82
Huyer, S. 85

Ibero-American Manual of Engagement Indicators (Valencia Manual) 227–228, 231
Ibero-American Network of Science and Technology Indicators (RICYT) 221, 221–222, 226, 227
ideology and values *187*, 188
impact of research
and citation counts 96
as criterion in PASRES 142
and debates around excellence 65, 92–93
as defining part of quality/RE 61
and evaluation of scientific knowledge production 204
and measurement of 261
and narrow focus on 73–74, 95
and non-academic impact 178, 179, 197
and Southern perceptions of funders' priorities 176n3
and Southern views on RE 168, *168*, 169, 170, 173, 175
in sustainable development context 208–209
of think tanks *187*, 188

implementation research 169–170
inclusivity in research 11, 33–34, 63, 88–89, 174, *253*
India *125*, 173, 248
indigenous knowledge/research 22, 62, 97, 108
  see also locally based research
Indonesia
  Jokowi's R&D plan 119–120
  and New Public Management 121–123
  and research excellence 131–135
  RIRN research plan 123, 124, *124*, 126–127
  S&T assessment and measurement 123–131
  S&T policy overview 120–121
innovation systems
  and engagement between researchers and society 49
  influence on science policy 48
  Rome Declaration on Responsible Research and Innovation in Europe 77n3
  see also science, technology and innovation (ST&I)
Institute for Fiscal Studies (UK) 190, 195
integrity of research 214, 251, *253*, 255
international collaboration
  and assessment of quality/RE 72, 135
  and issues around autonomy 44–45, 45
  and levels of social capital 82
  see also team-based research
International Development Research Centre (IDRC)
  funding of SGCI 43
  funding to Makerere University *158*
  meta-analysis of research projects 254–255
  see also Research Quality Plus model
international funders 44–45, 46, 48, 52
  see also funding of research
international research systems 99, *102*, 107
Israel 163

journal articles
  gender disparities in 85–86
  PASRES support to researchers 144
  recognised as significant *ex post* achievement 61
  'salami' publications 8, 36
  and searches for 235, 239
  views of Indonesian academics on 130
  see also language, and research reporting; peer-review processes; writing
journals
  in Côte d'Ivoire 143–144
  as gatekeepers of quality research 3, 205, 236
  Journal Impact Factor of 235–236, 238, 239
  lack of respect for local publications 21
  predatory journals 7–8
  public policies in support of 225
  and publication of interdisciplinary research 171
  see also open access publishing; publishers and publishing

Kang, A.J. 85–86
Kenya 184, 241
Kraemer-Mbula, E. 39, 63, 68, 73, 75, 83, 164–165
Krefting, L.A. 82

language, and research reporting 3, 21, 72, 103, 109, 170, 171–173, 175, 250

see also journal articles; writing
Latin America
bibliographic databases 224
and concept of 'peripheral science' 26
loss of local talent 102–103
open access publishing 110, 110–111, 111, 236, 240, 263
overview of S&T 221–222
and Sabato-Botano triangle 23
ST&I indicators 28
support for scientific journals 225
surveys on R&D activities 224, *225*
think tanks 180, 194
universities' role in knowledge production 94, 226–230
see also names of countries
Laudel, G. 66
Leahey, E. 82
lecturers
and gender disparities 85
in Indonesia 128–131
Makerere University 153, 155
legacy publishers 233–235, 236–238, 241, 243
see also open access publishing; publishers and publishing
legitimacy of research 11, 214, 251, *253*, 255
Leiden Manifesto for Research Metrics 8, 10, 71–72, 256
literature reviews of RE 65–69
locally based research
and alignment with global developments 52
and appropriation of Northern knowledge 104–105
de-legitimisation of 22
and evaluation of 250
funding of 'applied science' 50
and Leiden Manifesto principles 71
and loss of local talent 101–103
privileging the interconnectedness of 106–107
publication of 21, 241
and relevance of research 9, 11, 40, 71, 74, 110, 173, 263
representation in databases 224
and research excellence 70, 97
and Sabato-Botano triangle 99, 100, *102*
and social knowledge creation 108, 110
value of international exposure 29
visibility of researchers 40, 72, 98
see also Global South; low- and middle-income (developing) countries (LMICs)
Lorde, A. 105
low- and middle-income (developing) countries (LMICs)
and bibliometric indicators 222, 224
as burgeoning research talent pool 260
funding of research 62–64, 157
and granularity of research 72
lack of autonomy at national level 44–46
low morale of researchers 29
poor performance in S&T indicators 28–29
and relevance of research 9
and research excellence 20, 27, 60–62, 70, 73, 76, 220
and research obstacles 3, 49–50
and role of researchers in international publications 67
see also Global South; locally based research
Lundvall, B-A. 49

Makerere University (Uganda)
overview of 147–151

funding from development partners 157–161
repositioning to a research-led university 151–153
research productivity 153–157
see also Uganda; universities; university rankings
Matthew Effect
and assessment of proposals 31
and competition for funds 29, 30
and debates around excellence 7
detection of 37
and support for early career researchers 32
measurement of quality 3, 217n1, 219–220, 261
see also quality of research
measurement of RE *see* assessment/ measurement of RE
Medie, P.A. 85–86
merit, scientific 168, *168*, 169, *174*, 176, 214, 251
Merton, Robert 64–65, 77n5, 205
meta-analysis of research projects 254–255
Mexico
bibliographic databases 224
National System of Researchers 19, 20, 21
R&D expenditure 221
research evaluation system 210–211
Moher, D. 75
Mouton, J. 84, 203

Narayanamurti, V. 50
National Council for Higher Education (Uganda) 151–152, 153–154
National Research Foundation (SA) 43
National System of Researchers (NSR)
in Mexico 19, 20, 21
in Uruguay 34–36
neo-colonial agenda of RE 93, 96–99, 104
see also colonialism; decolonisation; post-colonial countries
Netherlands 110, 111, *158*
networks
and assessment of RE 65, 72
and credibility of research *187*, 196
and funding of research 51–52
and gender disparities 82, 87–88
Ibero-American Network of Science and Technology Indicators (RICYT) 221, 221–222, 226, 227
as relevant attribute in assessment of proposals 27
and Sabato-Botano triangle 99–103, 108
SciELO network 224, 236, 240
and 'silo mentality' 67–68
New Public Management (Indonesia) 121–123
see also Indonesia
Nobel prize 7, 59
see also prizes and awards
non-profit publishers 236
North Africa 84
Northern research *see* Global North

O'Donnell, Guillermo 26
Odumosu, T. 50
Ofir, Z. 68, 70, 253
open access publishing
AAS Open Research platform 41, 241–243, 263
in framework of RRI 62
in Global South context 110–111
growing support for 236, 239–241
indicators of quality and importance 243–245
see also journal articles; journals; publishers and publishing

organisations
as unit of achievement 64
value of branding 191
value of gender diversity 79–80
original research
component of RQ+ model 251, *253*
as manifestation of RE 61, 139, 167, *168*
publication of 233, 235, 238–239, 243, *244*
as relevant attribute in assessment 27
and Southern perceptions of funders' priorities 176n3
outcomes of research
and debates around RE 92–93
as key indicator of RE 62, 74
and socially relevant research 70
outputs of research
and debates around RE 61, 62, 260
making a 'value' calculation of 237
narrow focus on 73–74
and NPM approach 121–122
and Southern perceptions of funders' priorities 176n3
and Southern views on RE 169, *169*
and work of think tanks 184, 197
see also assessment/measurement of RE; bibliometric measures; citation counts; publication counts/output; quantitative indicators

participation principle 209, 212, 213, 214–215
see also engagement in research
partnerships *see* international collaboration; team-based research
PASRES programme (Côte d'Ivoire)
overview of 138–140
activities around RE 144–145
selection and evaluation of projects 140–144
Pasteur's Quadrant 165, 166, *166*, 171
Pearce, Nick 192
peer-review processes
and bias in 3, 234
and *ex post* qualitative evaluations 7
and exercise of academic autonomy 41, 43, 95
in Indonesia 134, 135
and limitations of dominant evaluation approaches 249
at Makerere University 156
as mechanism to judge research 40, 66, 73, 205, 223, 225
in open access publishing 235, 240, 242–243, *242*, 244–245, *244*
in PASRES 143, 144
as primary role of SGCs 47
and Southern views on RE 168
and wastage of scholarly output 237
see also assessment/measurement of RE; evaluation of research; journal articles
'Perspectives on research excellence in the Global South' (Tijssen & Kraemer-Mbula) 63, 72–73
Peru 184, 256
PhD students/graduates
African Doctoral Dissertation Research Fellowship programme 253
and gender disparities 80
Makerere University 151, 152, *152*, 153, *154*, 155, 156, 158
in think tanks 184
see also universities
plagiarism *244*
pluralism
and concept of quality 107, 260
and debates around excellence 5, 11, 25, 262
in research evaluation processes 32–34, 35, 36, 37–38

Polanyi (Republic of Science) 41, 43, 44, 47
political capital 45, 47, 48
see also capital, forms of
post-colonial countries 93, 94, 97, 98, 103, 104, 106, 110, 111
see also colonialism; decolonisation; neo-colonial agenda of RE
predatory journals 7–8
see also journal articles; journals; publishers and publishing
Price, Derek de Solla 205
private sector
in evaluation/execution of research 143, 145
think tanks 182
prizes and awards 7, 59, 67, 145, 237, 239
professors
in Indonesia 129
at Makerere University 155, 156–157
women 80, 85
Programme Appui Stratégique à la Recherche Scientifique see PASRES programme (Côte d'Ivoire)
promotions
gender disparities in 79, 85, 87
linked to journal publications 21, 234, 237, 239
at Makerere University 155, 156
proposals *see* research proposals
Prozesky, H. 84
public engagement 191–194, 196–197
see also engagement in research
public–private partnerships 52, 53
public sector 50, 53, 62
publication counts/output
and bias towards established researchers 173
as criterion for evaluation 40, 61, 66
and debates around excellence 6, 7, 39
fears of misuse of 71
and gender issues 80–81, 82
as indicator in RES 204, 205–206, 212, 215
in Indonesia 128–129, 130–131, *133*
and language of researchers 172, *172*
at Makerere University 151–152, *152*, *154*, 155, 156, 159–160
in Mexico 19
and open access publishing 244
and RQ+ model 255
and Southern views on RE 167, *168*, 170
as Uruguayan NSR criterion 34–35, 36
see also assessment/measurement of RE; bibliometric measures; citation counts; outputs of research; quantitative indicators
publishers and publishing
barriers in Global South 233, 241, 250
changing landscape of 235–239
legacy publishers 233–235, 236–238, 241
loyalty to traditional publishing 238
see also journal articles; journals; open access publishing

quality of research
and consequences of research 164
in Côte d'Ivoire 139, 140
dominance of Northern standards 21, 167
Enhancement of the Quality of Research programme (Uruguay) 32
in Global South context 214, 254
implications for use-inspired research 170–171, 173
in Indonesia 123, 131, 132, 134, 135
and Journal Impact Factor 235–236, 238, 239

journals as gatekeepers of 3, 205, 236
judged by peer-review process 223
measurement of 3, 217n1, 219–220, 261
multidimensional nature of 5, 164–165, 214, 257
and open access publishing 243–245
pluralistic view of 5, 107, 260
and RE, concept of 54, 59, 61, 65
and RE, encompassed by *174*
and RE, fundamentally different from 260
and RE, recommendations for assessment of 73
and 'relevance' criterion 214
Southern views on 169, 173, 175–176
as tenet of RQ+ model 251, *253*
in Uganda 156, 162–163
and work of think tanks 185, 186, 187, *187*, 189–190, 196
see also research excellence (RE); Research Quality Plus (RQ+) model

quantitative indicators
and debates around excellence 6–9
deeply embedded in academic practices 96
narrow focus on 73–74, 74
scientometrics 77n5, 204, 205
see also assessment/measurement of RE; bibliometric measures; citation counts; outputs of research; publication counts/output

questions, research 168, 169, *169*, 174

rankings
and debates around excellence 3, 5, 7, 9, 67, 96
via production/citation indicators 205–206
see also university rankings

relevance of research
and autonomy of research 41, 42, 46, 47
component of RQ+ model 251, *253*
and engagement in research 51–52, 53
examples of solving societal challenges 248–249
and inclusion of 'non-scientists' 42, 48–49
in Indonesia 132, 134
as key parameter for funders 169
in local contexts 9, 11, 40, 71, 74, 110, 173, 263
in PASRES programme 139, 140, 143
and publication of research 223
and RE, as criterion in determining 40, *174*
and RE, debates around 8, 9, 11, 40, 65, 70, 93
and RE, reconciling with 40–41
and RE, Southern views on 168, *168*, 170, 174, 175, 176n4
and SGCs 50–51, 54
in sustainable development context 208–209, 213, 214, 215, 216
and work of think tanks 184, 185, *187*

Republic of Science (Polanyi) 41, 43, 44, 47

research agendas
alignment of local/global agendas 52, 54
and need for autonomy 43, 45, 45–46, 46, 54, *187*
value of pluralistic vision of research 11

research and development (R&D)
funding of 123, *125*, 221, 224, *225*
and higher levels of investment in 161
in Indonesia 119–120, 121

in Latin America 28, 221, 224, *225*
and priorities of international partnerships 45
in SSA 224, *225*
in sustainable development context 209
Research and Innovation Oriented Toward Social Inclusion (Uruguay) 32–34
Research Assessment Exercise (UK) 20, 21
research design 169, *169*, 176n3, 176n4
research environment 68, 251, *253*
research evaluation systems (RES)
contribution to economic growth 206
in sustainable development context 208–209, 210–216
use of quantitative indicators 204, 205–206, 212, 215
see also assessment/measurement of RE; evaluation of research
Research Excellence Framework (REF) (UK) 2–3, 9, 49, 208
research excellence (RE)
and definitions/understandings of 1–2, 59–60, 64–65, 69, 77n1, 92–94, 138
achieving consensus on 37–38, 75
call to action for transformation of concept 259–263
debates around 1–10
developmental view on 23–28
as methodological/political construct 173, *174*
multidimensional nature of 5, 61, 62, 63, 73
as neo-colonial agenda 93–94, 96–99, 104
review of academic studies 65–69
speculative history of 94–96
structuring effects of 20–23
zero-sum/threshold excellence 69–70, 75
see also assessment/measurement of RE; embedded excellence; quality of research
research grants 67, 84, 139, 176, 239
research master plan (RIRN) (Indonesia) 123, 124, *124*, 126–127
see also Indonesia
research methods
in think tanks 185, 188–189, 197
in use-inspired research 174
research policy
and conflicts around appraisal of merits 36–37
frame 1 ST&I policy 206
proposals to 'do good' through 28–29
research proposals
assessment by University Research Council (Uruguay) 31, 32–34
and debates around excellence 5
and relevant attributes in assessment of 27
and responsible assessment of RE 61
Research Quality Plus model (IDRC)
as alternative method of evaluating quality 10, 70, 249, 257
concerns about 255
road test of 252–253
in sustainable development context 214–215
three tenets of 251–252
see also International Development Research Centre (IDRC)
research questions 168, 169, *169*, 174
researchers
and barriers to publishing 233–234, 235, 239
bias towards 3, 173
capacity and motivations of 12–13
competition between 7–8
and debates around excellence 4–5, 5

and endorsement from other scientists 249
and establishment of research priorities 223
in Indonesia 124–125, *126*
in LMICs 29, 67
numbers per million people 162–163
percentages of world researchers 224, *225*
as unit of assessment 65
see also gender equality
resources
access to 82
distribution of 3, 12
scarcity of 11, 45, 84, 157, 246, 261, 263
Responsible Research and Innovation (RRI) 62, 77n3
Rich, Andrew 188
RICYT (Ibero-American Network of Science and Technology Indicators) 221, 221–222, 226, 227
rigour
and community engagement/ responsiveness 188–189
correlation with capacity building 254
promoted in UK's REF 2
and research excellence 61, 65, 139, 167, 173
and research quality *174*, 176, 251
and role of SGCs 47, 54
and Southern perceptions of funders' priorities 176n3
and usefulness of research 254–255
and work of think tanks 190
RIRN research plan (Indonesia) 123, 124, *124*, 126–127
see also Indonesia
Rome Declaration on Responsible Research and Innovation in Europe 62, 77n3
see also innovation systems; science, technology and innovation (ST&I)
Rothmann, S. 84

Sabato-Botana Triangle 22–23, *24*, 99–103, 108
'salami' publications 8, 36
see also publishers and publishing
San Francisco Declaration on Research Assessment 8, 256
Schmidt, E. 66
Schwandt, T. 70
Schwartz, John 191
SciELO network 224, 236, 240
science
and changing landscape of publishing 235–241
concept of undone science 21–22
Merton's views on 77n5, 205
and need to answer fundamental questions 26
perceptions of 'good science' 1
SAB's views on roles of 207–208
scientific merit 168, *168*, 169, *174*, 176, 214, 251
stereotypes based on gender 81
Science and Techno Parks (Indonesia) 127
see also Indonesia
science and technology (S&T)
in Indonesia 120–121, 123–131
and loss of local talent 101–103
and Sabato-Botano triangle 22–23, *24*, *100*, *101*, *102*
Uganda's 'Initiative for Science Support' 157
Science Citation Reports 96
Science Granting Councils Initiative (SGCI) 39, 43, 48, 263
science granting councils (SGCs)
definition of 39, 43

capability-building efforts 50–51
and debates around excellence 11–12
and funding issues 44–45, 46, 48, 49, 50, 51–52, 52, 53
and importance of research relevance 169
and need for engagement with ST&I actors 43
and need for participation/ stakeholder engagement 51–53
need for variety of 'capitals' 41, 45, 47, 48, 51
recommendations for 54–55
and Southern views on RE 175, 176
tensions between autonomy and engagement/embeddedness 41, 42–48
value of threshold excellence approaches 70
science, technology and innovation (ST&I)
and indicators for research evaluation 205–206
initiatives to support transformative innovation 46
in Latin America/LMICs 28–29, 49
in sustainable development context 207–210
see also innovation systems
science, technology and society (STS) 44
Scientific Advisory Board (UN) 207–208
scientific merit 169, *174*, 176, 214, 251
scientometrics 77n5, 204, 205
Scopus database 129, 159, *160*, 223, 224, *225*, 227, 239
serendipity 43
sexual harassment 248
Singh, S. et al. *166*, *168*, *169*, 171–172, *172*, *174*, 175–176
Soapbox (communications firm) 191
social capital 45, 47, 48, 51, 82, 173
social knowledge creation 107–110
Sørensen, M. 67
South Africa
challenges of de-colonisation 93
funding of SGCs 43, 49
gender disparities in research 83–84, 85
open access publishing 111
see also Global South; Sub-Saharan Africa (SSA)
South Korea 124, *125*
Southern research *see* Global South
Spain 171–172, *172*, 210–211, 227
stakeholder engagement 51–53, 167, *168*, 176n3
see also engagement in research
standardisation 3, 95, 170
stereotypes, gender 81
Stirling, Andrew 44
Stokes, Donald 165–166, *166*
Stone, Diane 180
Sub-Saharan Africa (SSA)
criteria in determining RE 39–40
HE enrolment rates 150, 161
surveys on R&D activities 224, *225*
see also Global South; low- and middle-income (developing) countries (LMICs); science granting councils (SGCs)
sustainable development
definitions of 206–207
and challenges to traditional scientometric indicators 207–210
and funding for innovation-related research 49
and funding for Makerere University 160
gender dimension to 79, 83
as goal in PASRES programme 142
and research evaluation systems 208–209, 210–216

and social context of research 43–44, 204
Sweden 49, 158, *158*, 163

team-based research
and achieving research excellence 11, 67–68
giving visibility to research 29
and issues around autonomy 44–45, 45
prevalent in Global South 72
and quality of Southern research 254
value of stakeholder engagement 52–53
see also international collaboration
tenure 81, 82, 129, 239
think tanks
overview of 180–182
and credibility of 185–190, 197–198
interaction with universities 179, 181–182
opportunities for research and communication 190–197
and research excellence 183–185
threshold excellence 69, 69–70, 75
see also embedded excellence; research excellence (RE)
Tijssen, R. 39, 63, 68, 73, 75, 83, 164–165
transformation principle 208–209, 213, 214–215, 216
Transformative Innovation Policy Consortium (TIPC) 53
Transformative Innovation Policy (TIP) 208
transparency
as aspect of good research practice 105
as *Leiden Manifesto* principle 71
in open access publishing *242*, 243, *244*
promoted in UK's REF 2
and research excellence 63, 74, 261
in work of think tanks *187*, 196, 197

Uganda
higher education sector 148, 148–149, 150, 151–152, 153–154, 161, 162
number of universities 148
see also Makerere University (Uganda)
undone science 21–22
see also entries under science
United Kingdom
development of RES 204
and 'impact agenda' 95
initiatives in Indonesia 123
open access publishing 110
Research Assessment Exercise 20, 21
Research Excellence Framework 2–3, 9, 49, 208
think tanks 190, 192, 195
university rankings 19, 20
United Nations (UN)
Scientific Advisory Board 207–208
and sustainable development 207, 210
United States
funding to Makerere University *158*
projects in Indonesia 122–123
think tanks 188, 192
universities
African universities 83–84, 161, 162
in colonial settings 94
developmental 24–25, 83
and gender disparities in research performance 83–84
in Indonesia 122, 125, 126, 128–131, 135
interaction with think tanks 179, 181–182

key in supporting economic growth/ innovation 226
in Uganda 148, 152
see also higher education sector; Makerere University (Uganda)
University of the Republic (Uruguay) 30
university rankings
and aims of the university 24
deeply embedded in academic practices 96
and homogenisation of assessment 74
in Indonesia 135
and RE, debates around 6, 9
and RE, drive for 19–20
university research centres 196–197, 198
Uruguay
implementation of NSR 34–36
R&D/ST&I indicators 28
University Research Council 30–34
use-inspired research 165–166, *166*, 167, 170, 170–171, 171, 173, 174, 175
usefulness of research 119, 168, 184, 206, 254–255
users of research 8, 11–12, 61, 68, *189*, 197, 243–244, 251

Valencia Manual (RICYT) 227–228, 231
validation of research 7, 98, 240, 249
values and ideology *187*, 188
verification of research 189, 190, 196, *244*
visibility of research and researchers
challenges in Global South 11, 72
gender disparities in 84
increased through research teams 29
and Matthew Effect 7
in open access publishing 242, 244, *244*
in think tanks *187*, 188, 196
via publication/citation counts 35

Web of Science (WOS) database 223, 224, 239
Wellcome Open Research *240*
Wellcome Trust 55n8, 263
Western Balkans 182, 192
Widodo, President Joko (Jokowi) 119–120
see also Indonesia
Williams, W.M. 82
women professors 80, 85
see also gender equality
World Bank 122
writing
academic vs popular articles 130
and language of researchers 171–173
within PASRES programme 143, 144–145
and work of think tanks 184, 186, 192
see also journal articles; language, and research reporting
Wynne, Brian 44

Young, John 185
Young, M. 69

zero-sum excellence 69, 70, 75
see also embedded excellence; research excellence (RE)

www.ingramcontent.com/pod-product-compliance
Ingram Content Group UK Ltd.
Pitfield, Milton Keynes, MK11 3LW, UK
UKHW020202290926
15471UKWH00005B/1026

9 781928 502067